MOSFET CIRCUITS GUIDEBOOK— with 100 Tested Projects

By Rufus P. Turner

TAB BOOKS
Blue Ridge Summit, Pa. 17214

FIRST EDITION

FIRST PRINTING—OCTOBER 1975
SECOND PRINTING—DECEMBER 1977
THRID PRINTING—JUNE 1978

Printed in the United States
of America

Hardbound Edition: International Standard Book No. 0-8306-5796-7

Paperbound Edition: International Standard Book No. 0-8306-4796-1

Library of Congress Card Number: 75-27483

Contents

Preface 9

Chapter 1
Meet The MOSFET 11
Description of MOSFET—Operating Modes—Advantages and Disadvantages—Gate-Protected MOSFET—MOSFETS Used in This Book—Hints and Precautions

Chapter 2
AF Amplifiers 18
Single-Stage RC-Coupled Amplifier—High-Gain Single-Stage Amplifier—Two-Stage High-Gain Amplifier—Degenerative Amplifier—Source Follower—MOSFET—Bipolar Amplifier with AGC—MOSFET Input For IC Amplifier—Dual-Input AF Mixer—Dual-MOSFET Phase Inverter—Paraphase Amplifier—Gated-On Amplifier—Gated-Off Amplifier—LC-Tuned Low-Pass Amplifier—RC-Tuned Low-Pass Amplifier—LC-Tuned High-Pass Amplifier—RC-Tuned High-Pass Amplifier—LC-Tuned Peak Amplifier—RC-Tuned Peak Amplifier—LC-Tuned Notch Amplifier—RC-Tuned Notch Amplifier—Headphone Amplifier—Peaked-Response Headphone Amplifier

Chapter 3
DC, RF, And IF Amplifiers 60
DC Amplifier—DC Voltage Amplifier—DC Galvanometer Amplifier—DC Source Follower—General-Purpose RF Amplifier—Single-Ended Transmitter-Type RF Amplifier—Push-Pull Transmitter-Type RF Amplifier—100 MHz RF Amplifier—Single-Tuned IF Amplifier—Double-Tuned IF Amplifier—Regenerative IF Amplifier—Video (Wideband) Amplifier

Chapter 4
Control Circuits and Devices 81
Zero-Current DC Relay—AC RF Relay—Sensitive AF Relay—LC-Tuned AF Relay—RC-Tuned AF Relay—Coincidence Relay—Touch-Plate Relay—Capacitance Relay—Sound Switch—Interval Timer—Continuously Variable Phase Shifter—Step-Type Phase Shifter—Sensitive Carrier-Failure Alarm—Logic AND Circuit—Logic OR Circuit—DC Signal Inverter—DC Impedance Converter—Dual-Polarity Output Adapter

Chapter 5
Oscillators 116
Tickler-Type Audio Oscillator—Hartley Audio Oscillator—Colpitts Audio Oscillator—Franklin Audio Oscillator—Phase-Shift Audio Oscillator—Wien-Bridge Audio Oscillator—Villard Audio Oscillator—Multivibrator—Sun-Powered Audio Oscillator—Self-Excited RF Tickler-Coil Oscillator—Self-Excited RF Hartley Oscillator—Self-Excited RF Colpitts Oscillator—Standard Crystal Oscillator—Pierce Crystal Oscillator—Sun-Powered RF Oscillator—Self-Excited IF Oscillator—Crystal-Controlled IF Oscillators

Chapter 6
Test Instruments 145
Single-MOSFET Electronic DC Voltmeter—Balanced Electronic DC Voltmeter—Center-Zero Electronic DC Voltmeter—Self-Excited Frequency Standard—AF–RF Signal Tracer—Tuned RF—IF Signal Tracer—AF Signal-Tracer Adapter For AC Voltmeter—Active Probe—Wide-Range LC Checker—AC Null Detector—Sensitive Light Meter—Sine- to Square-Wave Converter—AM Monitor—CW Monitor

Chapter 7
Miscellaneous Circuits 170
Constant-Current Adapter—Charge Detector (Electroscope)—Flip-Flop—Supersonic Pickup—Balanced Modulator—Simple Light-Beam Receiver—Sensitive Light-Beam Receiver—Regenerative Broadcast Receiver—All-Wave Regenerative Receiver—RF Mixer—Tuning Meter—Heterodyne Eliminator—Flea-Power "QRP" Transmitter—Frequency Doubler—Reactance Modulator

Index 194

Preface

The metal-oxide-semiconductor field-effect transistor, abbreviated MOSFET, is easily applied to electronic circuits of various types. In several respects, the MOSFET resembles the vacuum tube more closely than do the bipolar transistor and the conventional field-effect transistor, and this feature enables some circuits to be more easily transistorized.

This book offers a collection of tested MOSFET circuits, many of them employing only one transistor. These are largely old vacuum-tube favorites of experimenters and hobbyists, adapted to the MOSFET. Very little space is devoted to MOSFET theory; the reader is assumed to know already how a regular field-effect transistor works and to need only a brief fill-in on special features of the MOSFET.

For convenience, the circuits are grouped into conventional categories: amplifiers, oscillators, instruments, and so on. Each chapter is completely self-contained, so that the reader is not forced to refer to prior sections. The selected circuits are practical, and they offer an excellent opportunity to get acquainted with the MOSFET.

General instructions for handling, installing, and using MOSFETS of the type shown in this book are given in Chapter 1 under the heading *Hints and Precautions*, and the reader should study these carefully before starting his experiments.

Rufus P. Turner

Chapter 1

Meet The MOSFET

This chapter describes the principal features of the MOSFET. Since the reader is assumed to know how a conventional field-effect transistor works and is structured, no space is devoted here to basic theory. Only those MOSFET features are discussed that the reader must be familiar with in order to use this device safely and effectively. For additional material, both theoretical and practical, the reader is referred to the considerable literature on field-effect devices.

The instructions for handling and using the MOSFET should be read carefully by newcomers to MOSFET practice before they experiment with the device.

DESCRIPTION OF MOSFET

The MOSFET (metal-oxide-semiconductor field-effect transistor) is a special type of field-effect transistor in which the gate electrode is a small metal plate insulated from the substrate by a thin film of silicon dioxide. The gate leakage current—as low as 10 picoamperes (10^{-11} ampere) in some models—is much less than that in the junction FET and results in an input resistance comparable to that of the vacuum tube. To all practical intents and purposes, therefore, the MOSFET, like the vacuum tube, is essentially a voltage-actuated device.

Figure 1-1 shows the basic structure of an N-channel MOSFET. Although this may not be the precise cross section of a *particular* unit, it illustrates the main differences between MOSFETs and conventional junction FETs. In this sketch, the various regions are not to scale. In the P-type silicon wafer

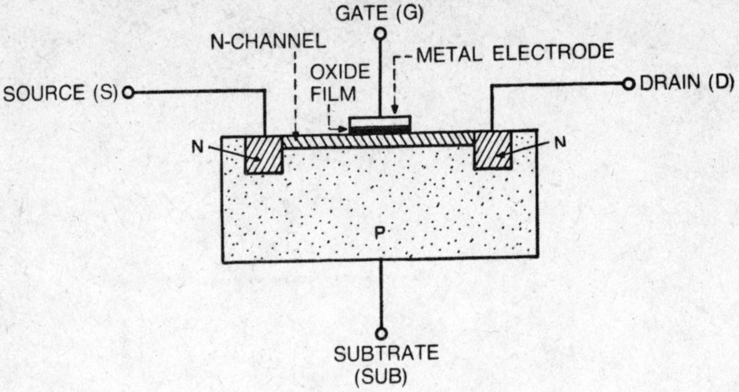

Fig. 1-1. Basic MOSFET structure.

(the substrate), the source (*S*) and drain (*D*) electrodes consist of small N-regions that are processed into the wafer. When an external voltage is applied between the source and drain, current carriers flow through the "channel" between these two regions. The gate electrode (*G*), which electrostatically controls these carriers, consists of a small metal film placed close to, but not touching, the wafer. The thin oxide film provides the insulation needed to keep the metal gate out of contact with the wafer. The corresponding circuit symbol is shown in Fig. 1-2A. This particular kind of device is known as an N-channel MOSFET.

If the Ns and Ps are interchanged in Fig. 1-1, we have instead a P-channel MOSFET; the circuit symbol will differ only in reversal of the arrow. For the N-channel device, the external DC voltages are: drain positive, source negative. For the P-channel device, the voltages are: drain negative, source positive.

Single-gate MOSFETs and dual-gate MOSFETs both are commercially available. Type 3N128, for example, is a single-gate unit, and type 3N140 is a dual-gate unit. Figure 1-2B shows the circuit symbol of a dual-gate MOSFET. Note that there are two gates and one drain; thus, the dual-gate MOSFET is comparable to a tube such as type 6AE7-GT, which has two control grids and one plate. The dual-gate device is particularly useful in converter, mixer, push-push doubler, and AGC-controlled circuits.

OPERATING MODES

There are three operating modes for MOSFETs: *depletion* mode, *enhancement* mode, and *depletion—enhancement* mode. In the depletion mode, a negative gate bias is employed, and a negative signal voltage applied to the gate narrows the channel and reduces the drain current. In the enhancement mode, a positive gate bias is employed, and a positive signal voltage widens the channel and increases the drain current. In the depletion–enhancement mode, zero gate bias is employed, and a positive signal voltage increases drain current (enhancement), whereas a negative signal voltage decreases drain current (depletion). These polarities apply to N-channel MOSFETs, and the opposite polarities apply to P-channel MOSFETs.

ADVANTAGES AND DISADVANTAGES

In many applications, the MOSFET offers advantages over the conventional FET. Chief among these advantages is very high input resistance, since the MOSFET gate is insulated from the semiconductor, the only gate current is the exceedingly tiny dielectric leakage—between 10 pA and 50 nA, (10 pA = 10^{-11}A, 50 nA = 5×10^{-8}A), depending upon make and model. In some devices, this resistance may be more than 100 million megohms. Very-high-resistance MOSFETs thus offer virtually no load.

Other advantages are: (1) lower noise (2 to 6 dB typical), (2) reduced cross-modulation effects, (3) low feedback capacitance (0.02 to 0.17 pF typical), (4) high maximum operating frequency (up to 500 MHz, depending upon make and model), and (5) high transconductance (up to 15,000 micromhos typical).

The only pronounced disadvantage of the MOSFET is the delicateness of its gate insulation. The thin silicon-dioxide film may easily be punctured beyond repair by electrostatic charges or induced voltages when handling the device, or by transients or overvoltage when using it. Extreme care is required in the handling, installation, and operation of

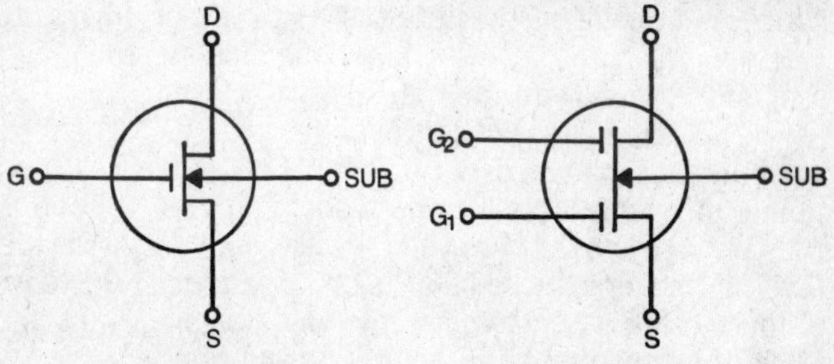

(A) SINGLE-GATE (B) DUAL-GATE

Fig. 1-2. MOSFET symbols.

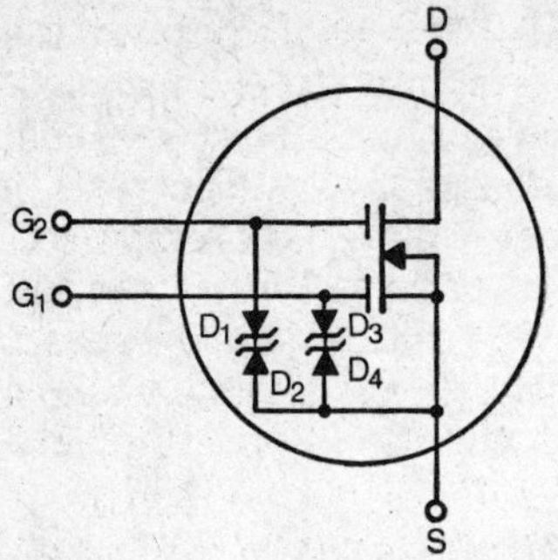

Fig. 1-3. Gate-protected MOSFET.

conventional MOSFETs. However, a special *gate-protected* MOSFET is commercially available (as a dual-gate model), and this type can be handled with the ordinary care given to any transistor. This nearly foolproof MOSFET is described below.

Some appraisers cite the input capacitance of the MOSFET—due to the equivalent capacitor formed by the metal gate, oxide insulation, and underlying semiconductor wafer—as a disadvantage. However, this capacitance (typically 3 to 6.5 pF) is no great impediment to the designer or user, as it is only slightly higher than the input capacitance of many tubes having the same transconductance, and is very much smaller than the input capacitance of some tubes.

GATE-PROTECTED MOSFET

Figure 1-3 shows the circuit symbol of the gate-protected MOSFET. In this arrangement, each gate is internally connected through a pair of integrated back-to-back-connected zener diodes (D_1 and D_2 for gate 2, and D_3 and D_4 for gate 1) to the substrate and source. At low voltages (including normal gate bias and normal peak signal voltage), the reverse-connected diode in each pair is essentially nonconducting; i.e., it offers extremely high resistance and does not interfere with MOSFET performance. (For a given polarity of gate voltage, one diode in each leg is reverse biased and the other is forward biased; when the polarity reverses, the bias conditions are reversed.) The diode leakage current is of the order of 1 to 50 nA (1 nA = 10^{-9}A), depending upon MOSFET make and model and gate voltage. In classic fashion, when the gate voltage rises above a prescribed safe level, the reverse-biased diodes undergo zener breakdown and conduct relatively heavily, providing a short-circuit path around the gate and protecting the gate insulation.

The integrated diodes do increase the gate leakage current and the input capacitance of the gate-protected MOSFET somewhat, but in many applications the protection against MOSFET destruction is well worth this slight disadvantage.

MOSFET USED IN THIS BOOK

In the experimenter's interest, only gate-protected MOSFETs are used in the circuits given in this book. Moreover, for simplicity and because of its wide range of usefulness, a single type of MOSFET—the RCA 3N187—is employed throughout. (If the reader desires, the RCA 40673 consumer-type MOSFET may be substituted for the 3N187.)

The exact circuit symbol of the 3N187 is shown in Fig. 1-4A. This sketch shows the internal elements, as well as the lead arrangement (G_1 = gate 1, G_2 = gate 2, S = source, D = drain). Figure 1-4B shows a simplified symbol for the 3N187. In this latter symbol, which is used in all the circuits in this book, the internal diodes have been omitted for simplicity (just remember that you are using a diode-protected MOSFET, and you do not need to make any connections to these diodes). Figure 1-4C shows the lead arrangement in the 3N187 MOSFET.

The 3N187 is an N-channel depletion-type MOSFET which may be operated as high as 300 MHz. It is enclosed in a TO-72 metal can; it thus is only approximately 0.21 in. high and 0.195 in. in diameter. Its four stiff leads may be plugged into a socket (e.g., Sylvania ECG-418) or soldered or welded directly into a circuit.

The following pertinent electrical characteristics of the 3N187 are taken directly from the manufacturer's literature.

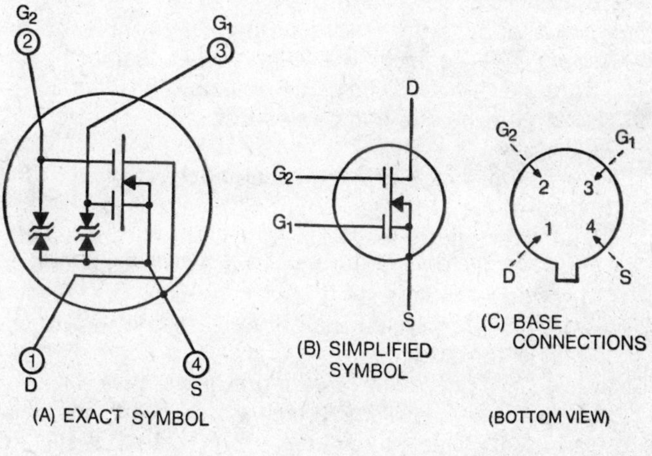

Fig. 1-4. 3N187 symbols.

Drain-to-source volts (V_{DS})	−0.2V to +20V max
Gate-1-to-source volts (V_{G1S})	+3V to −6V DC, ±6V peak AC
Gate-2-to-source volts (V_{G2S})	−6% to 30% of V_{DS} DC
Maximum drain current (I_D)	50 mA
Device dissipation (25°C)	330 mW
Power gain (G_{ps})	18 dB typ at 200 MHz
Noise figure (NF)	4.5 dB max at 200 MHz
Forward transconductance (G_{fs})	7000 μmho min; 12,000 μmho typ
Gate leakage current (I_{GSS})	50 nA max
Input capacitance (C_{iss})	6 pF typ
Reverse transfer capacitance (C_{rss})	0.03 pF max
Gate-1-to-source pinchoff voltage (V_{PG1})	−2V typ
Gate-2-to-source pinchoff voltage (V_{PG2})	−2V typ
Output resistance $\frac{(r_{oss})}{(V_{DS})} = 15V$, $I_D = 10$ mA, $V_{G2S} = 4V$, $f = 200$ MHz	2800Ω

HINTS AND PRECAUTIONS

For maximum protection of the MOSFET and best results from its use, the following hints are offered:

1. Employ the exact values of components and voltages specified in the circuit diagrams. Adjust a circuit (where this is needed) exactly as instructed in the text.
2. The reader is free to use his favorite method of construction: perforated board, open chassis, metal box, printed circuit. Use the same techniques and precautions that would apply to any other transistorized device.
3. All methods of transistor mounting are suitable for the MOSFET. These include use of socket, soldering or welding directly into circuit (use a suitable heatsink), use of clips, use of mounting screws, use of terminal studs.
4. After completing a project, check all wiring carefully, and insert the MOSFET *last*.
5. Be sure the pigtail leads are straight before inserting the MOSFET into a socket, and insure that each lead is in the correct hole (see Fig. 1-4C). To drive the MOSFET home, push firmly, but gently and straight down, on the top of the case.
6. Like other transistors, a gate-protected MOSFET such as the 3N187 is reasonably rugged mechanically and need not be handled gingerly. Nevertheless, it should not be abused mechanically. Avoid rough handling, dropping, hammering, and similar abuse. Do not

stress the pigtail leads. Use care when removing a MOSFET from its socket.

7. The metal case of the 3N187 is "hot"; that is, the case is internally connected to the substrate and source. Therefore, do not allow the case to come into contact with leads, chassis, or other components.
8. In all MOSFET circuits, use the shortest and most direct leads that are practicable. This will minimize stray pickup, undesired coupling, and undesired feedback.
9. When using only one gate of a dual-gate MOSFET, do not allow the other gate to float, especially if a lead is connected to it. Either ground the unused gate, connect it to the source, or connect a 1000Ω resistor between this gate and ground. A floating gate is highly susceptible to stray pickup and body capacitance.
10. In all RF circuits and high-gain AF circuits, shield all susceptible parts of the circuit just as you would in a tube circuit. When long leads are unavoidable, lead dress is important.
11. Do not subject the MOSFET to excessive currents or voltages. (Refer to the electrical characteristics of the 3N187.)
12. Never allow the combined gate voltage (DC bias plus peak AC signal voltage) to exceed the maximum gate-voltage rating of the MOSFET.
13. Avoid exposing the MOSFET to strong magnetic fields.
14. Protect the MOSFET from excessive heat.
15. In some of the circuit diagrams, a dashed line runs to the ground symbol. This means that the connection to chassis or to earth, as the case may be, is optional and depends upon how the circuit will be used by the reader. When, instead, a solid line runs to the ground symbol, the connection *must* be made.

Chapter 2

AF Amplifiers

This chapter presents 22 audio-frequency circuits, including single-stage amplifiers, 2-stage amplifiers, tuned amplifiers, and phase inverters. One circuit shows a common way of using a MOSFET with a bipolar transistor to obtain the best features of each; another circuit shows how a MOSFET may be operated ahead of a conventional integrated circuit to provide high input impedance for the latter. Some newcomers to MOSFET applications will choose these audio applications as a means of getting acquainted with the device.

In each of the circuits, unless otherwise indicated on the circuit diagram or in the text, capacitances are in picofarads and resistances in ohms. Resistors are ½W, and electrolytic capacitors have a DC working-voltage rating of 25V. The similarity of the components to those used in equivalent tube circuits will be readily apparent to the experienced reader.

For simplicity, batteries are shown for DC supply; however, a *well filtered* power supply may be used instead.

Before undertaking the wiring and operation of any circuit, read carefully the hints and precautions given in Chapter 1.

SINGLE-STAGE RC-COUPLED AMPLIFIER

Figure 2-1 shows the circuit of a single-stage amplifier employing a 3N187 MOSFET in the common-source circuit. This circuit is equivalent to the grounded-cathode tube circuit and the common-emitter bipolar-transistor circuit.

The MOSFET is biased to operate in its linear region by means of a combination of (1) gate-1 negative bias developed

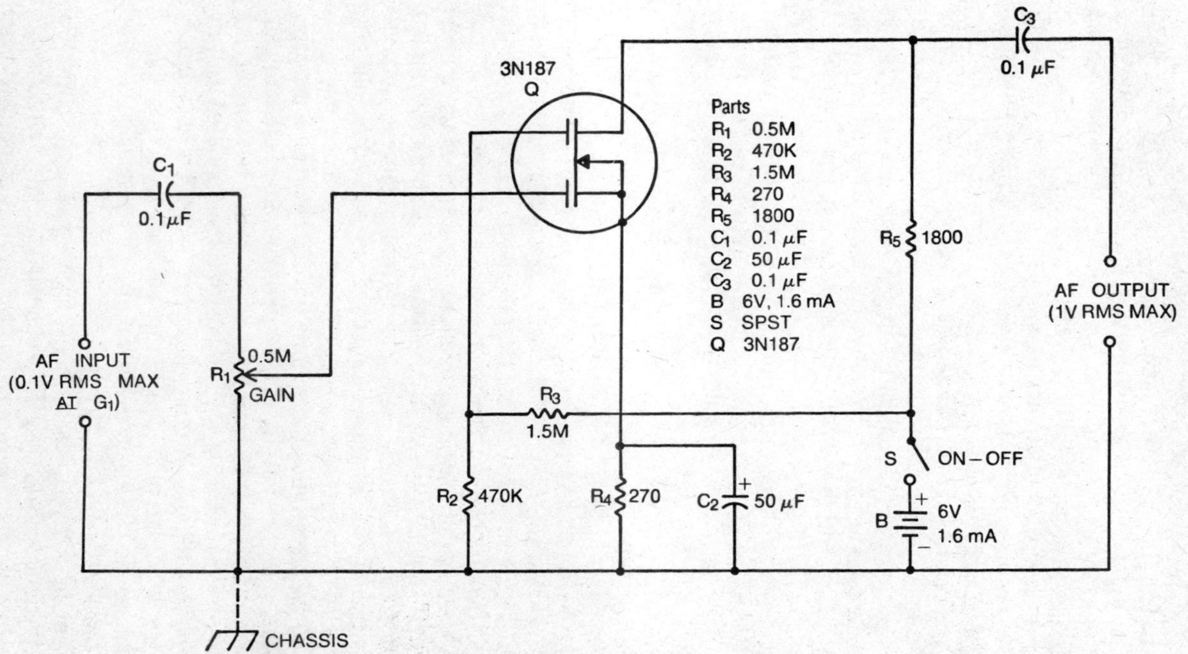

Fig. 2-1. Single-stage RC-coupled amplifier.

by the flow of drain current through source resistor R_4, and (2) gate-2 positive bias developed by voltage divider R_2-R_3. In this divider, resistance R_3 may need to be adjusted with an individual 3N187 for linear operation.

The input resistance of the amplifier is approximately ½M, the resistance of the gain-control potentiometer R_1. The high gate resistance of the MOSFET enables the use of a high-resistance potentiometer here.

Current drain from the 6V supply B is approximately 1.6 mA. Open-circuit voltage gain of the single stage is 10. The maximum input-signal voltage before output-peak clipping is 0.1V RMS at gate 1 (higher amplitude signals are reduced to this maximum by appropriate settings of potentiometer R_1). The corresponding maximum output is 1V RMS. Response of the circuit is flat within ±2 dB of its 1000 Hz response from 50 Hz to 25 kHz. If the DC supply is increased to 12V (2.6 mA), the open-circuit voltage gain increases to 15 (maximum input-signal voltage equals 0.2V RMS at gate 1; corresponding maximum output-signal voltage before peak clipping equals 3V RMS).

HIGH-GAIN SINGLE-STAGE AMPLIFIER

Operating with a higher DC voltage and different circuit constants, the single-stage *RC*-coupled amplifier shown in Fig. 2-2 provides almost four times the voltage gain afforded by the similar amplifier described in the preceding section. There are numerous applications in which high single-stage gain is desired: preamplifiers, AC voltmeter amplifiers, signal boosters for electronic relays and other control devices, electronic games, oscilloscope amplifiers and preamplifiers, etc.

The amplifier is biased into its linear region by a negative voltage applied to gate 1 and a positive voltage applied to gate 2. The negative bias is developed by the flow of drain current through source resistor R_4; the positive bias is developed by the voltage divider R_2-R_3, operated from the DC supply (battery B). In an individual setup, resistor R_3 in this divider may need some adjustment for linear operation of the amplifier.

The input resistance of the amplifier is approximately ½M, the full resistance of the gain-control potentiometer R_1 Higher input resistance may be obtained, at some chance of stray pickup, by increasing the resistance of this potentiometer. The high gate resistance of the MOSFET enables the use of a high-resistance potentiometer in this position.

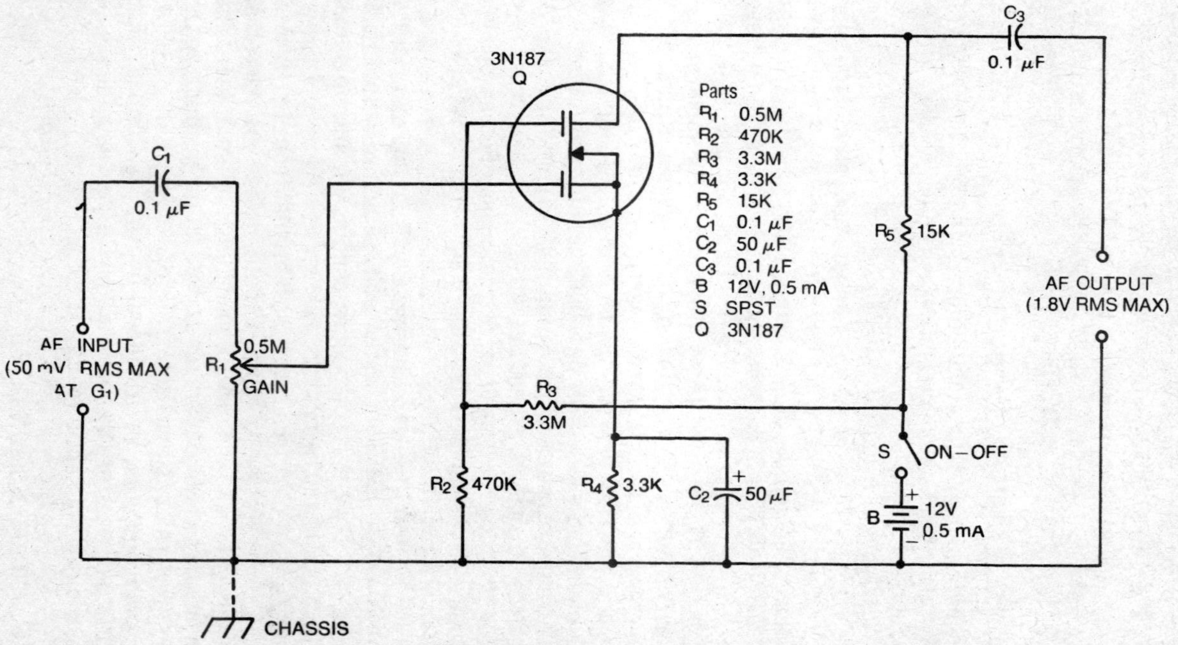

Fig. 2-2. High-gain single-stage amplifier.

Current drain from the 12V supply B is approximately 0.5 mA. Open-circuit voltage gain of the single stage is 36. The maximum input-signal voltage before output peak clipping is 0.05V RMS at gate 1 (higher amplitude signals are reduced to this maximum by appropriate settings of potentiometer R_1). The corresponding maximum output is 1.8V RMS. Response of the circuit is flat within ±2 dB of its 1000 Hz response from 100 Hz to 50 kHz, and is −5 dB at 20 Hz.

TWO-STAGE HIGH-GAIN AMPLIFIER

Figure 2-3 shows the circuit of an amplifier employing two 3N187 MOSFETs in cascade. The open-circuit voltage gain of this amplifier is 1200.

The stages are individually biased, with gate 1 of each MOSFET receiving a negative voltage, and gate 2 of each MOSFET a positive voltage. In the first stage, gate 1 of MOSFET Q_1 receives the voltage drop developed across source resistor R_4 by the flow of drain current, and gate 2 receives the voltage developed by voltage divider R_2–R_3. In the second stage, gate 1 of MOSFET Q_2 receives the voltage drop developed across source resistor R_{10} by the flow of drain current, and gate 2 receives the voltage developed by voltage divider R_8–R_9. With individual MOSFETs, voltage-divider resistors R_3 and R_9 may require some adjustment for best linearity of the amplifier.

The input resistance of the amplifier is approximately 1M, determined by gate-to-ground resistor R_1. A higher resistance may be used for R_1, if desired, at some risk of stray pickup. The high gate resistance of the MOSFET Q_1 permits use of a high resistance for R_1.

Current drain from the 12V supply B is approximately 1 mA. The maximum input signal voltage before output peak clipping is 1.5 mV RMS. The corresponding maximum output signal voltage is 1.8V RMS.

Response of the circuit is reasonably flat from 100 Hz to 20 kHz. The response is approximately −2 dB down at frequencies lower than 100 Hz and higher than 20 kHz. The frequency response may be altered by suitably changing the capacitance of C_1, C_3, and C_6.

Stable operation is enhanced by the decoupling filter consisting of resistor R_6 and capacitor C_4. This coupler should not be omitted from the circuit.

DEGENERATIVE AMPLIFIER

The benefits of negative feedback in reducing distortion and generally improving amplifier operation are obtained in the single-stage amplifier shown in Fig. 2-4 by employing an

Parts			
R_1	1M	R_{11}	15K
R_2	470K	C_1	0.1 μF
R_3	3.3M	C_2	50 μF
R_4	3.3K	C_3	0.1 μF
R_5	15K	C_4	10 μF
R_6	1K	C_5	50 μF
R_7	0.5M	C_6	0.1 μF
R_8	470K	B	12V, 1 mA
R_9	3.3M	S	SPST
R_{10}	3.3K	Q_1	3N187
		Q_2	3N187

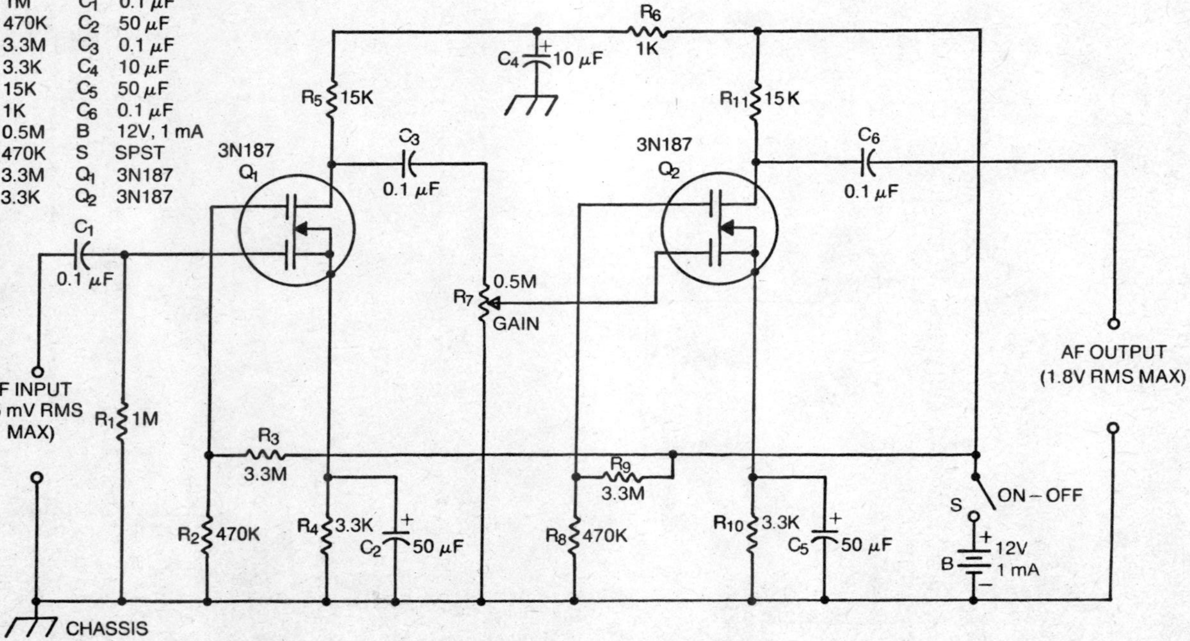

Fig. 2-3. Two-stage high-gain amplifier.

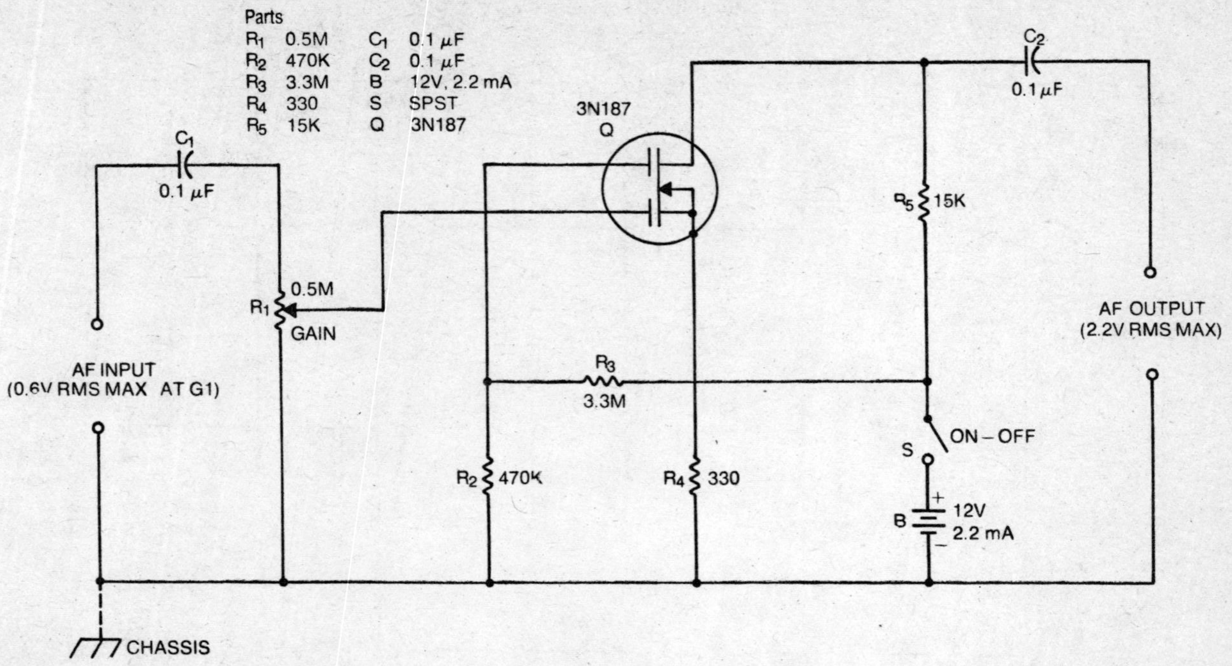

Fig. 2-4. Degenerative amplifier.

unbypassed source resistor R_4. The price for the improved operation is, of course, loss of gain. Thus, the voltage gain of this circuit, compared with that of its counterpart shown in Fig. 2-1, is approximately 3.7.

The amplifier is biased into its linear region by a negative voltage applied to gate 1 of MOSFET Q_1 and a positive voltage applied to gate 2. The negative bias is developed as the voltage drop across source resistor R_4 (produced by the flow of drain current through this resistor). The positive voltage is developed by voltage divider R_2–R_3. With an individual MOSFET, resistor R_3 in this divider may need some adjustment for best linearity of amplifier operation.

The input resistance of the amplifier is approximately ½M, determined principally by the resistance of the gain-control potentiometer R_1. The high gate resistance of the MOSFET permits use of this high resistance at R_1. If desired, a higher R_1 resistance may be employed, at some risk of stray pickup.

Current drain from the 12V supply is approximately 2.2 mA. The maximum input-signal voltage before output peak clipping is 0.6V RMS at gate 1 (higher amplitude signals are reduced to this maximum by appropriate settings of potentiometer R_1). The corresponding maximum output signal voltage is 2.2V RMS.

SOURCE FOLLOWER

Figure 2-5 shows the circuit of a simple source follower. This circuit is equivalent to the vacuum-tube cathode follower and the bipolar-transistor emitter follower. Like the latter two circuits, the source follower is a degenerative type employing negative feedback to cancel stage-introduced distortion. The circuit is characterized by high input impedance and low output impedance. As such, it has many well known applications in electronics, especially that of impedance transformer with power gain. The source follower, like the cathode follower and emitter follower, is noted also for its wide frequency response and low distortion.

In this circuit, gate 1 and gate 2 are connected together and receive negative DC bias from the voltage drop produced by the flow of drain current through source resistor R_2. The high bypass capacitance C_2 places the drain effectively at ground potential. The input resistance of the stage is approximately 1M and is largely determined by the full resistance of gain-control potentiometer R_1. For higher input resistance, when this is desired, the resistance of R_1 may be increased, at some risk of stray pickup. The approximate

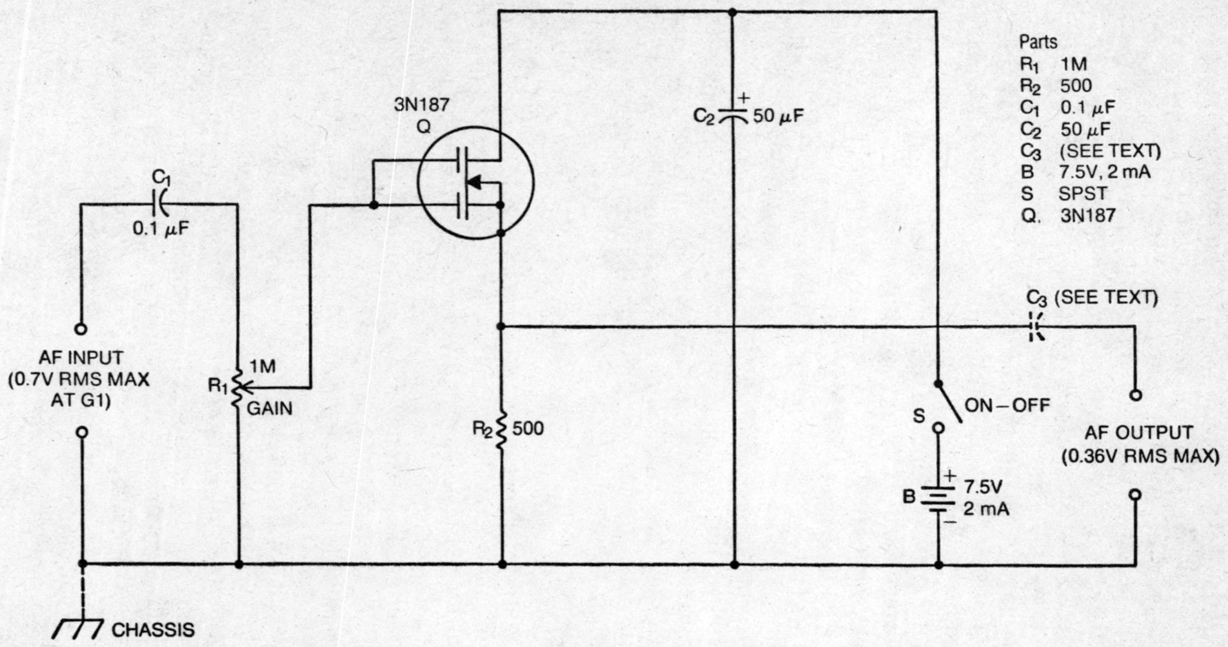

Fig. 2-5. Source follower.

effective output impedance of the stage is 277Ω. But this will differ somewhat with individual MOSFETS, since transconductance is a term in the impedance formula and varies with MOSFETS of the same type. The reader can vary R_2 for a desired value of output impedance. The output impedance is given by

$$R_o = \frac{r_{oss} R_2}{(Y_{fs} r_{oss} + 1) + r_{oss}} \qquad \text{(Eq. 2-1)}$$

where

R_o = output impedance (ohms)
R_2 = source resistor value (ohms)
r_{oss} = MOSFET output resistance (ohms). (See manufacturer's literature, but figure approximately 2800Ω for the 3N187.)
Y_{fs} = transconductance of the MOSFET (mhos) (See manufacturer's literature.)

Voltage gain of the stage is approximately 0.51. The maximum input signal voltage before output peak clipping is 0.7V RMS at the gates of the MOSFET (higher amplitude signals are reduced to this maximum by appropriate settings of potentiometer R_1). The corresponding maximum output signal voltage is 0.36V RMS. You may obtain higher output voltage by increasing resistance R_2, and vice versa, but this will also affect the value of output impedance (see Eq. 2-1). The equation for voltage gain of the source follower is

$$G_v = \frac{Y_{fs} R_2}{1 + Y_{fs} R_2} \qquad \text{(Eq. 2-2)}$$

where Y_{fs} and R_2 nave the same meanings as in Eq. 2-1. Response of the stage is reasonably flat from 100 Hz to 100 kHz and is approximately 2 dB down at 20 Hz. Current drain from the 7.5V supply is approximately 2 mA.

Output capacitor C_3 will be required when this source follower drives a stage or device to which direct coupling will be undesirable. Its capacitance will be governed by the resistance or inductance that will be encountered in the driven device and the extent to which the *LC* or *RC* combination alters the frequency response of the source follower.

MOSFET–BIPOLAR AMPLIFIER WITH AGC

Figure 2-6 shows the circuit of a 2-stage amplifier employing a 3N187 MOSFET in the input stage and a 2N2712 silicon bipolar transistor in the output stage. The stage

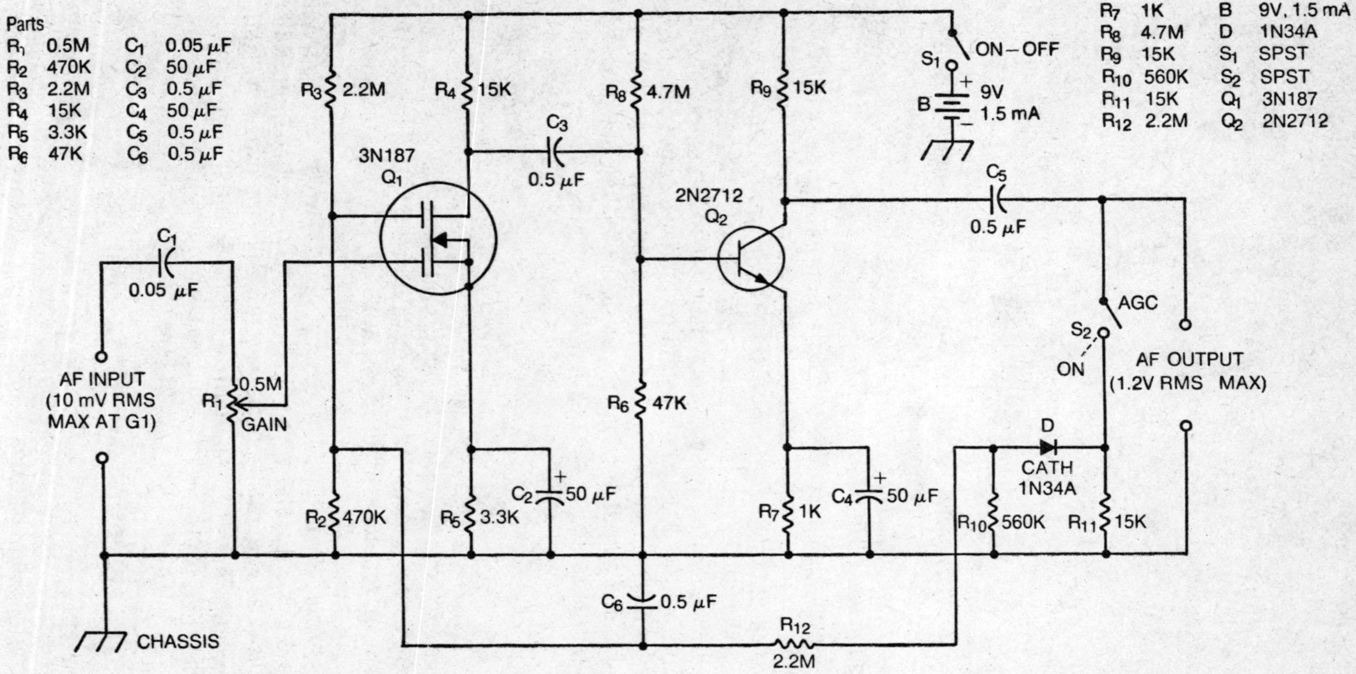

Fig. 2-6. MOSFET – bipolar amplifier with AGC.

configuration is similar: common-source MOSFET and common-emitter bipolar transistor. This arrangement affords the high input impedance of the MOSFET and the high voltage gain of the bipolar transistor. An added advantage is overall automatic gain control.

In the first stage, negative gate-1 bias for the MOSFET is provided by the voltage drop resulting from the flow of drain current through source resistor R_5, and positive gate-2 bias is supplied by voltage divider R_2–R_3 (for an individual MOSFET, resistor R_3 in this divider may require some adjustment for linear operation of the amplifier). In the second stage, the base bias of the 2N2712 transistor is the combination of the voltage drop resulting from the flow of collector current through emitter resistor R_7 and the output of voltage divider R_6–R_8.

The input impedance of the amplifier is approximately ½M, determined principally by the full resistance of gain-control potentiometer R_1. A higher resistance (e.g., 1M) may be used, if desired. The high gate resistance of the MOSFET permits use of a high-resistance gain control in what is essentially a transistorized amplifier.

Current drain from the 9V supply is approximately 1.5 mA. The overall voltage gain (into open circuit or very-high-resistance load) is 120. The maximum input signal voltage before output peak clipping is 10 mV RMS at gate 1 of MOSFET Q_1 (higher amplitude signals are reduced to this maximum by appropriate settings of potentiometer R_1). The corresponding maximum output signal voltage is 1.2V RMS. These figures are obtained with AGC switch S_2 open (i.e., AGC off). When S_2 is closed, the rectifier circuit D–R_{10}–R_{11} is cut into the circuit, rectifies a portion of the output signal, and sends the resulting DC through filter R_{12}–C_6 to gate 2 of the MOSFET Q_1. The germanium diode D is poled for negative DC output, and this negative voltage at gate 2 of Q_1 reduces the gain proportionately in the input stage.

MOSFET INPUT FOR IC AMPLIFIER

While integrated circuits having MOSFET transistors processed into them are obtainable, a great many conventional IC amplifiers with conventional bipolar transistor input are in the hands of hobbyists and experimenters. Figure 2-7 shows one way of providing high-impedance MOSFET input for such an integrated circuit. The IC here is an RCA 3020. This integrated circuit is a complete audio amplifier having three intermediate stages and a push-pull class B output stage delivering ¼W of AF power. The input resistance provided by the MOSFET Q is governed largely by the resistance of

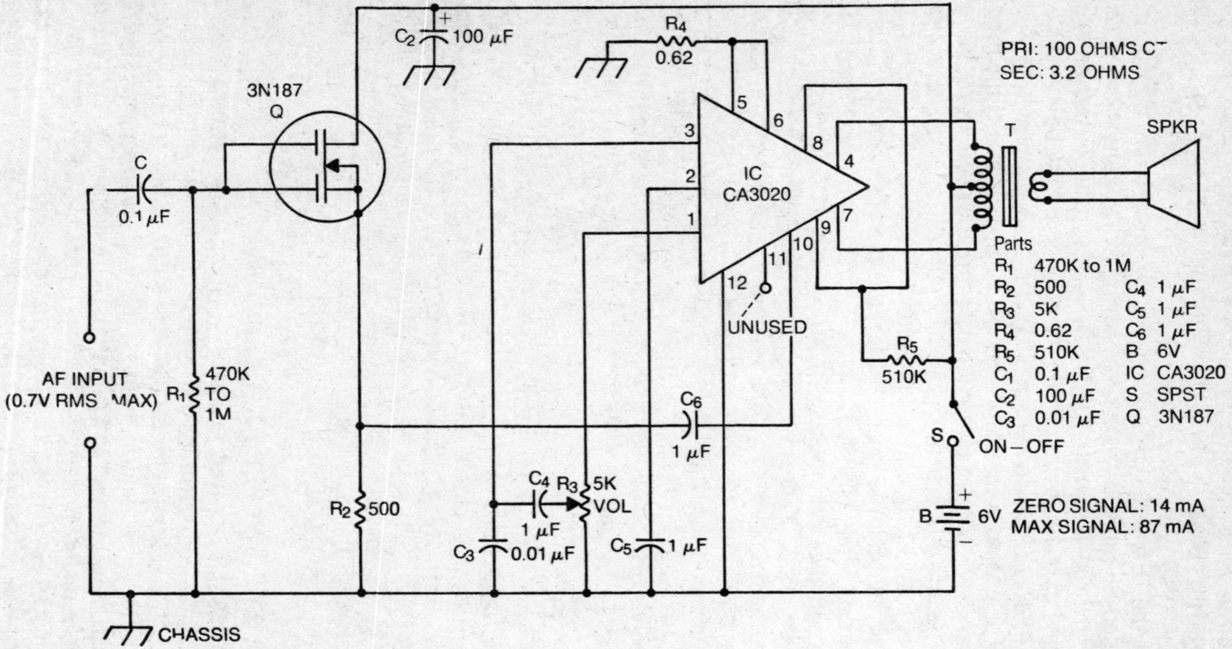

Fig. 2-7. MOSFET input for IC amplifier.

gate-to-ground resistor R_1 and can be any value between 470K and 1M. The higher values will make the setup more susceptible to stray pickup. The input stage is a source follower.

In the outboard input stage, the two gate electrodes of the MOSFET are connected together and receive their negative bias from the voltage drop developed across source resistor R_2. The output of the IC is coupled to a 3.2Ω speaker through a miniature, transistor-type transformer T having a 100Ω, centertapped primary winding and 3.2Ω secondary.

Since the IC contains a class B stage, the direct-current drain will be different under quiescent and driven conditions: The zero-signal current drain from the 6V source thus is 14 mA, and the maximum-signal drain is 87 mA, approximately. Higher audio power may be obtained by appropriately increasing the battery voltage; however, for powers in the vicinity of ½W, a heatsink must be used with the IC. For ¼W output with minimum distortion, the maximum input signal voltage is 0.7V RMS. Since the gain control R_3 is at the input of the second stage of this circuit, the signal must be limited to this maximum at the amplifier input.

DUAL-INPUT AF MIXER

One of the merits of the dual-gate MOSFET is its low cross-modulation characteristic. This feature allows the device to be employed effectively in converters and mixers. Figure 2-8 shows the circuit of a dual-input audio-frequency mixer employing a single 3N187 MOSFET.

Here, AF input 1 is applied to gate 1 of the MOSFET through ½M gain-control potentiometer R_1, and AF input 2 is applied to gate 2 through ½M gain-control potentiometer R_2. Gate 1 receives negative bias resulting from the voltage drop developed by the current through source resistor R_5 gate 2 receives positive bias produced by the output of voltage divider R_3–R_4. The common output signal is developed across drain resistor R_6 and is coupled to the output through capacitor C_5. The resistance at each signal input is approximately 0.5M, determined largely by the resistance of potentiometer R_1 or potentiometer R_2. Higher input resistance may be obtained by substituting higher resistance potentiometers, at some risk of stray pickup.

Current drain from the 6V source is approximately 3 mA. The open-circuit voltage gain for each half of the circuit is 10. The maximum input signal voltage before output peak clipping is 0.1V RMS at gate 1 or gate 2 (higher amplitude signals are reduced to this maximum by appropriate settings of

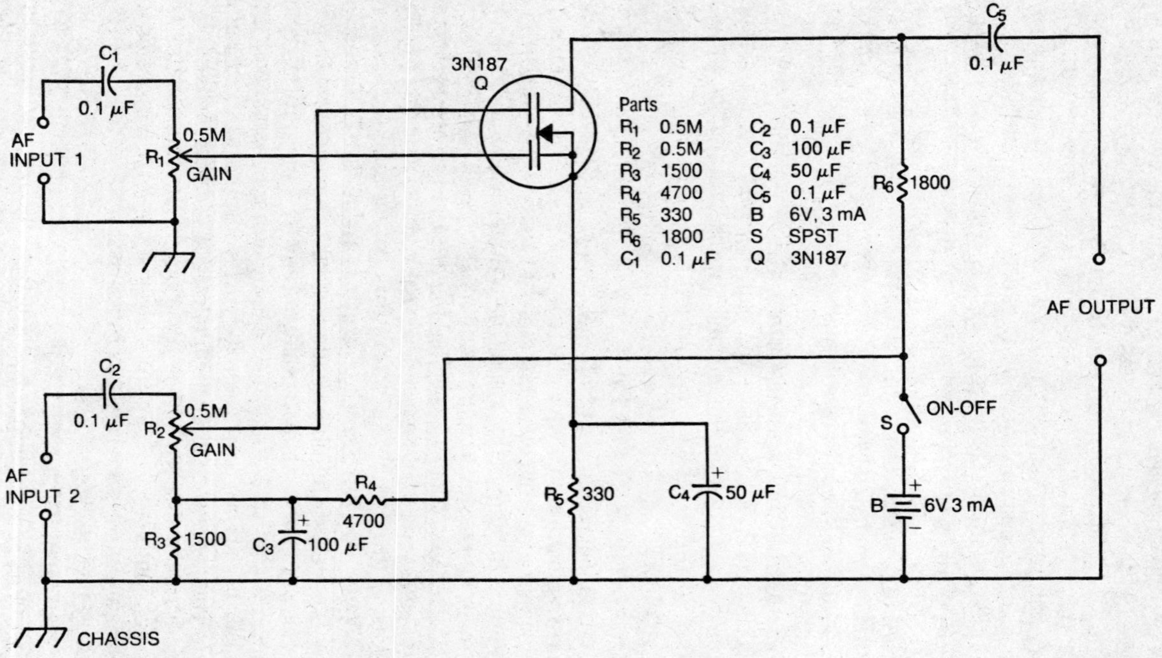

Fig. 2-8. Dual-input AF mixer.

gain-control potentiometers R_1 and R_2). The corresponding maximum output signal voltage is 1V RMS.

Like its tube and transistor counterparts, this mixer will operate from a combination of devices. Examples are: two microphones, two players, one microphone and one player, and so on.

DUAL-MOSFET PHASE INVERTER

Figure 2-9 shows the circuit of a conventional dual-device phase inverter adapted for MOSFETs. This circuit is applicable to many experimental uses, as well as to its usual application of driving a push-pull amplifier. In this arrangement, the AF input signal is first amplified by MOSFET Q_1, which, because of the common-source circuit, provides a 180° phase shift at the junction of R_9 and C_4. This output signal is sampled by potentiometer R_{11}, which returns a fraction of the signal to the input of MOSFET Q_2, which then amplifies this fraction and shifts its phase by 180°. The output of Q_2 is delivered to the phase-2 output terminal. In this way, the two output signals are of opposed phase, with respect to the common output terminal. Balance-control potentiometer R_{11} is adjusted for equal amplitude of the phase-1 and phase-2 output signals.

In this circuit, gate 1 of MOSFET Q_1 receives negative bias resulting from the voltage drop produced by the flow of drain current through source resistor R_4, and gate 2 of this MOSFET receives positive bias from voltage divider R_2–R_3. Resistor R_3 in this divider may need some adjustment with an individual MOSFET for linear operation of the top half of the circuit. Similarly, gate 1 of MOSFET Q_2 receives negative bias resulting from the flow of drain current through source resistor R_8, and gate 2 of this MOSFET receives positive bias from voltage divider R_6–R_7. Resistor R_7 of this divider may need some adjustment with an individual MOSFET for linear operation of the bottom half of the circuit.

Current drain from the 6V source is approximately 3.5 mA. The maximum input signal voltage before output peak clipping is 0.1V RMS. The corresponding output signal voltage is 1V RMS. The open-circuit voltage gain of each half of the circuit accordingly is 10. Excellent balance may be obtained through careful adjustment of potentiometer R_{11}.

PARAPHASE AMPLIFIER

When the demands for close balance are not so stringent, a paraphase inverter may be used in place of the 2-device phase inverter just described. A suitable paraphase-type circuit is

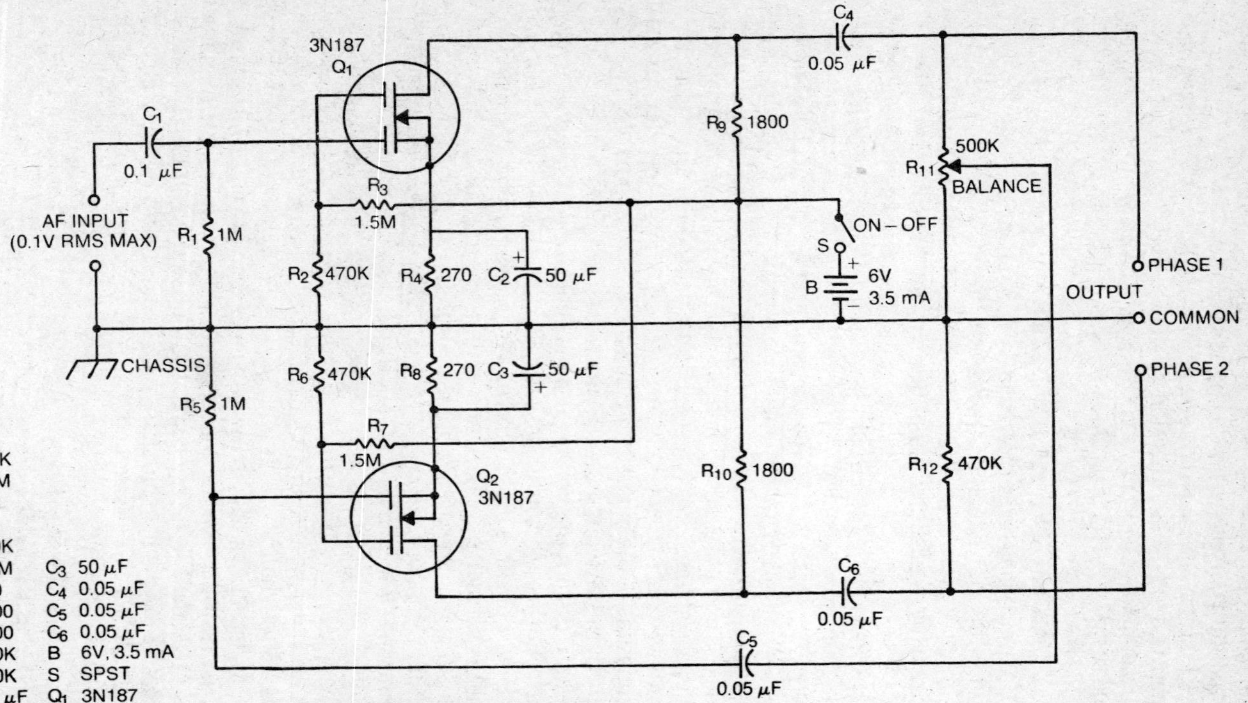

Parts

R_1	1M		
R_2	470K		
R_3	1.5M		
R_4	270		
R_5	1M		
R_6	470K		
R_7	1.5M	C_3	50 μF
R_8	270	C_4	0.05 μF
R_9	1800	C_5	0.05 μF
R_{10}	1800	C_6	0.05 μF
R_{11}	500K	B	6V, 3.5 mA
R_{12}	470K	S	SPST
C_1	0.1 μF	Q_1	3N187
C_2	50 μF	Q_2	3N187

Fig. 2-9. Dual-MOSFET phase inverter.

shown in Fig. 2-10. An obvious advantage of such a circuit is its use of just one active device.

In this arrangement, a single 3N187 MOSFET supplies each component of the output signal. Because this is a common-source circuit, the applied signal undergoes a 180° phase shift at the drain output (junction of R_2 and C_2) and no phase shift (0°) at the source output (junction of R_3 and C_3). In most instances, R_2 and R_3 will be identical; sometimes, however, a particular MOSFET will need adjustment of one of these resistances, usually R_3, for identical phase-1 and phase-2 output voltages.

Current drain from the 6V source is approximately 1.5 mA. The maximum input signal amplitude before output peak clipping is 0.4V RMS. The corresponding maximum output signal voltage is 0.3V RMS. The open-circuit voltage gain therefore is 0.75, somewhat like that of a source follower. The input resistance of the stage is largely the resistance of gate-to-ground resistor R_1. For most applications, 0.47M will be a satisfactory value for R_1; for higher input resistance, however, up to 10M may be employed, at some risk of stray pickup and pronounced hand-capacitance effects. If desired, fixed resistor R_1 may be replaced with a potentiometer.

GATED-ON AMPLIFIER

Figure 2-11 shows one type of circuit in which an amplifier is switched on by means of a gating signal; the amplifier automatically switches off, as far as signal transfer is concerned, when the gating signal passes. In this common-source circuit, negative bias is applied to gate 2 of the MOSFET by 1.5V source B_1 through the voltage divider R_3–R_4. This bias is sufficient to cut the MOSFET off, with normal negative bias supplied to gate 1 through the voltage drop across source resistor R_5. In the cutoff condition, the amplifier can pass no signal.

When subsequently a 3V trigger or gating voltage is applied to the trigger input terminals, enough voltage is developed by the R_2–R_3 voltage divider to buck the cutoff bias of gate 2 and allow the amplifier to operate. A signal then is transmitted between the amplifier input and output with an open-circuit voltage gain of 5. When the 3V potential is removed, the MOSFET resumes its cutoff condition, and the output disappears.

Current drain from the 12V source B_2 is approximately 0.5 mA. Current drain from the cutoff-bias source B_1 is approximately 2.8 mA. The maximum input signal voltage

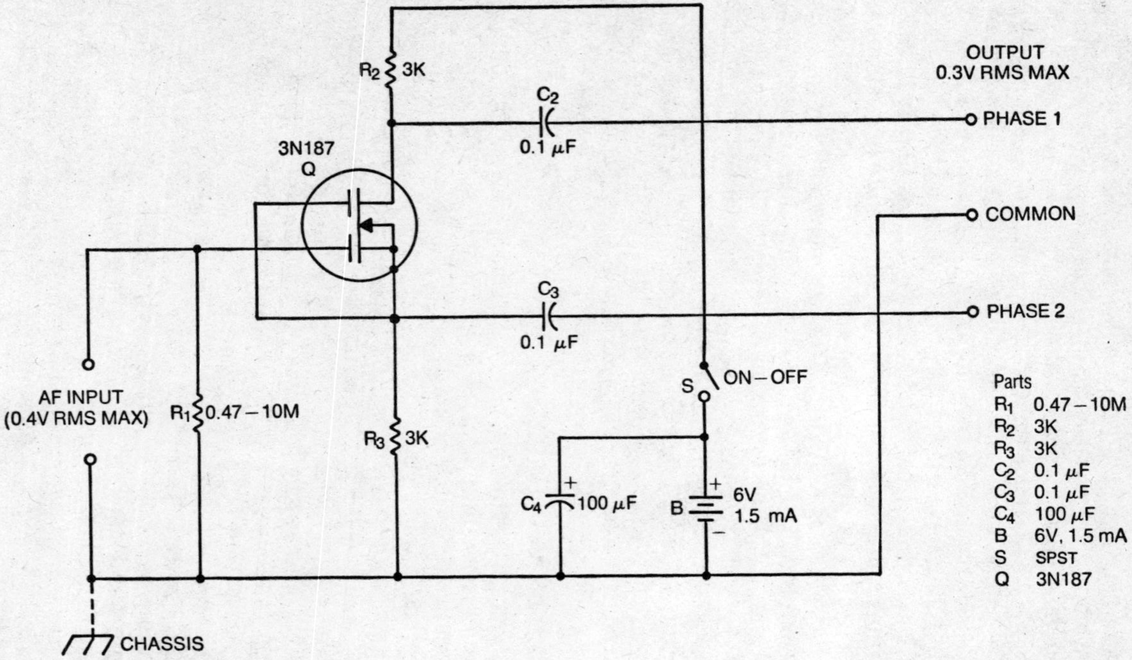

Fig. 2-10. Paraphase amplifier.

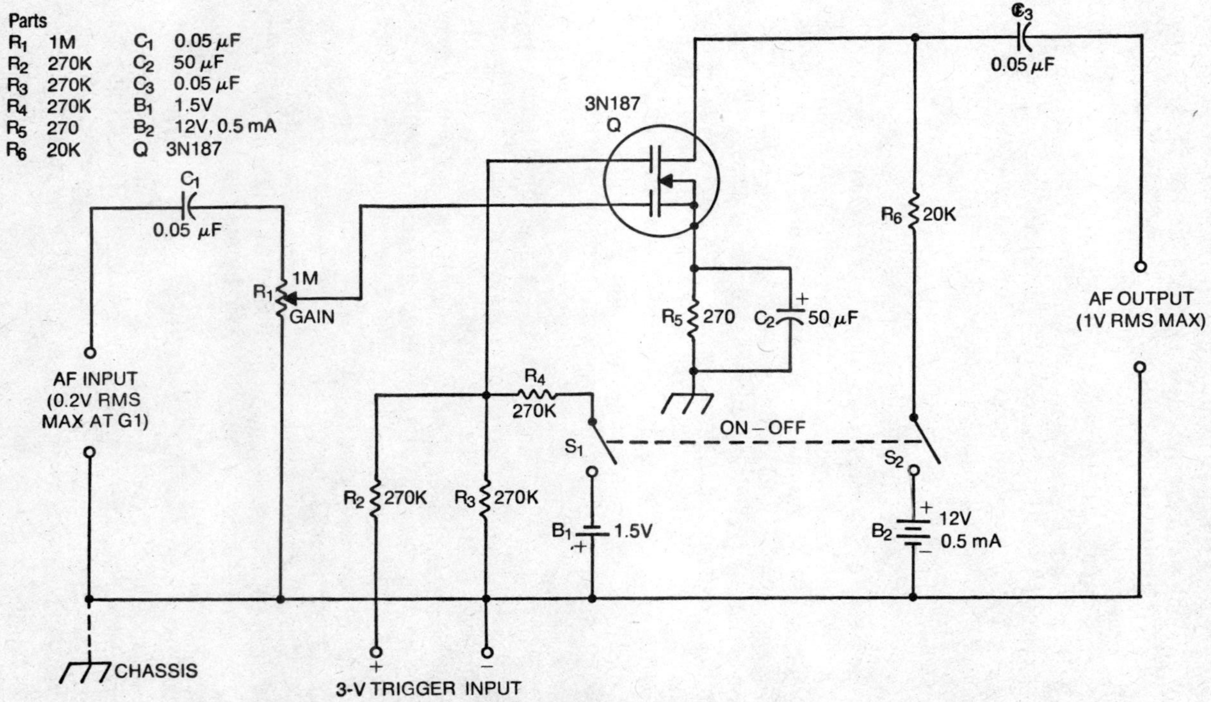

Fig. 2-11. Gated-on amplifier.

before output clipping is 0.2V RMS at gate 1 of the MOSFET (higher amplitude signals are reduced to this maximum by appropriate settings of gain-control potentiometer R_1). The corresponding maximum output signal voltage is 1V RMS. The input resistance of the circuit is approximately 1M, determined largely by the full resistance of potentiometer R_1; but this may be increased, if desired, by using a higher-resistance potentiometer. However, at the higher resistances, there is increased risk of stray pickup.

A gated-on amplifier has various applications in automatic control equipment and measuring instruments.

GATED-OFF AMPLIFIER

Figure 2-12 shows one type of circuit (the opposite of that described in the preceding section) in which an amplifier is switched off by means of a gating signal; the amplifier automatically switches on, as far as signal transfer is concerned, when the gating signal passes. In this common-source circuit, negative bias is applied to gate 2 of the MOSFET, by the gating voltage divider R_2-R_3, to pinch off the MOSFET. When this gating bias is absent, the MOSFET is normally biased by the negative voltage (resulting from the flow of drain current through source resistor R_4) applied to gate 1.

When no signal is applied to the trigger input terminals, the amplifier transmits a signal between its input and output with an open-circuit voltage gain of 5. When a 2V potential is applied to the trigger input terminals in the polarity shown here, however, the MOSFET is pinched off and the output signal vanishes. When the 2V gating signal is removed, the amplifier automatically resumes operation.

Current drain from the 12V source is approximately 0.5 mA. The maximum input signal voltage before output peak clipping is 0.2V RMS at gate 1 of the MOSFET (higher amplitude signals are reduced to this maximum by appropriate settings of gain-control potentiometer R_1). The corresponding maximum output signal voltage is 1V RMS. The input resistance of the circuit is approximately 1M, determined largely by the full resistance of potentiometer R_1; but this may be increased, if desired, by using a higher resistance potentiometer. However, at the higher resistance, there is increased risk of stray pickup.

Like the gated-on amplifier described in the preceding section, this gated-off amplifier has various applications in automatic control equipment and measuring instruments.

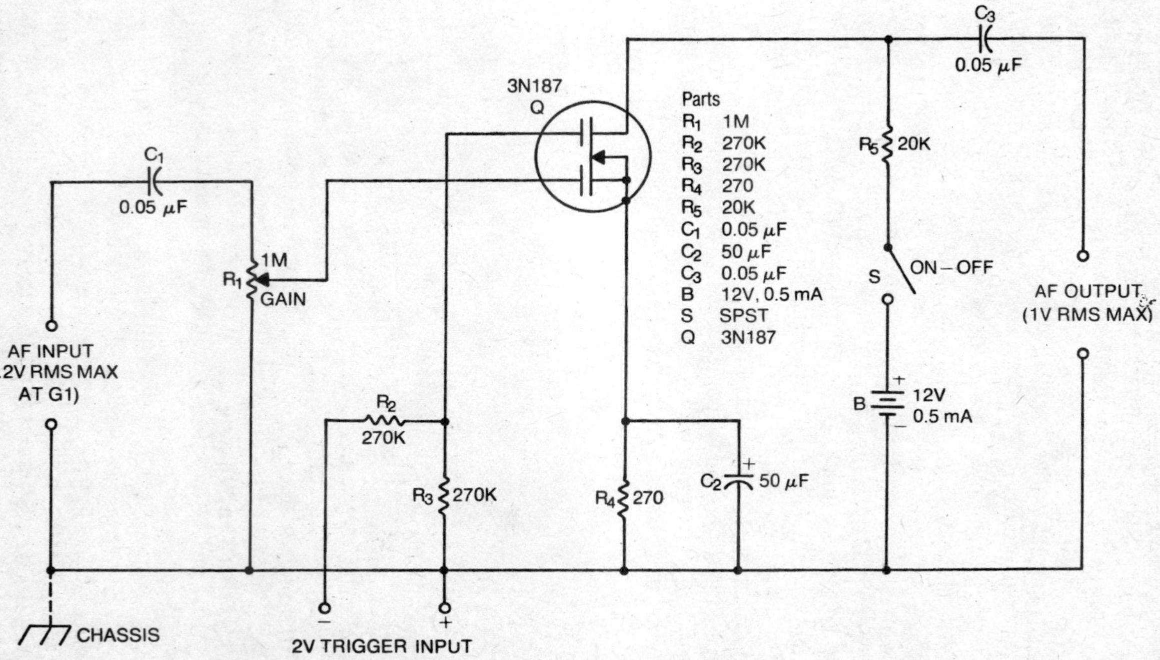

Fig. 2-12. Gated-off amplifier.

LC-TUNED LOW-PASS AMPLIFIER

Amplifiers may be tuned in various ways to pass certain frequencies and reject others. Figures 2-13 to 2-20 show circuits of such amplifiers. The advantages of tuned amplifiers over plain, passive tuned circuits or filters are (1) high input impedance, (2) voltage gain, (3) isolation of the tuning elements from external circuits, and (4) substitution in some instances of compact RC-tuned elements for bulkier inductors.

Figure 2-13A shows the circuit of an amplifier having the characteristics of a low-pass filter. Figure 2-13B shows the amplifier response; here, f_c is the cutoff frequency. This response is obtained by incorporating an *LC*-type *high*-pass filter section in the negative feedback loop between the drain output and gate-1 input circuit of the MOSFET, with the result that negative feedback occurs at all frequencies above f_c, canceling the gain at those frequencies, whereas the amplifier has approximately full gain below f_c.

Inductance L and capacitance C_2 are chosen for the desired cutoff frequency f_c according to the following equations, in which L is in henrys, C_2 in farads, and f_c in hertz:

$$L = \frac{10^4}{12.6f_c} \qquad \text{(Eq. 2-3)}$$

$$C_2 = \frac{1}{12.6f_c \times 10^4} \qquad \text{(Eq. 2-4)}$$

From Eq. 2-3, it is seen that for a cutoff frequency of 1000 Hz, $L = 794$ mH; and from Eq. 2-4, $C_2 = 0.0079\ \mu$F. Inductor L should be of as high quality as practicable to insure that it has high Q; otherwise, the filter rolloff will not be sharp.

In the basic circuit, MOSFET Q receives its negative gate-1 bias from the voltage drop resulting from the flow of drain current through source resistor R_5, and its positive gate-2 bias from voltage divider R_3–R_4. With an individual MOSFET, R_4 may need some adjustment for linear response of the amplifier. Current drain from the 6V supply is approximately 2 mA. Open-circuit voltage gain of the circuit is 10 in the passband. The maximum input signal voltage before output peak clipping is 0.1V RMS at gate 1 (higher amplitude signals are reduced to this maximum value by appropriate settings of potentiometer R_1). The corresponding maximum output signal voltage is 1V RMS.

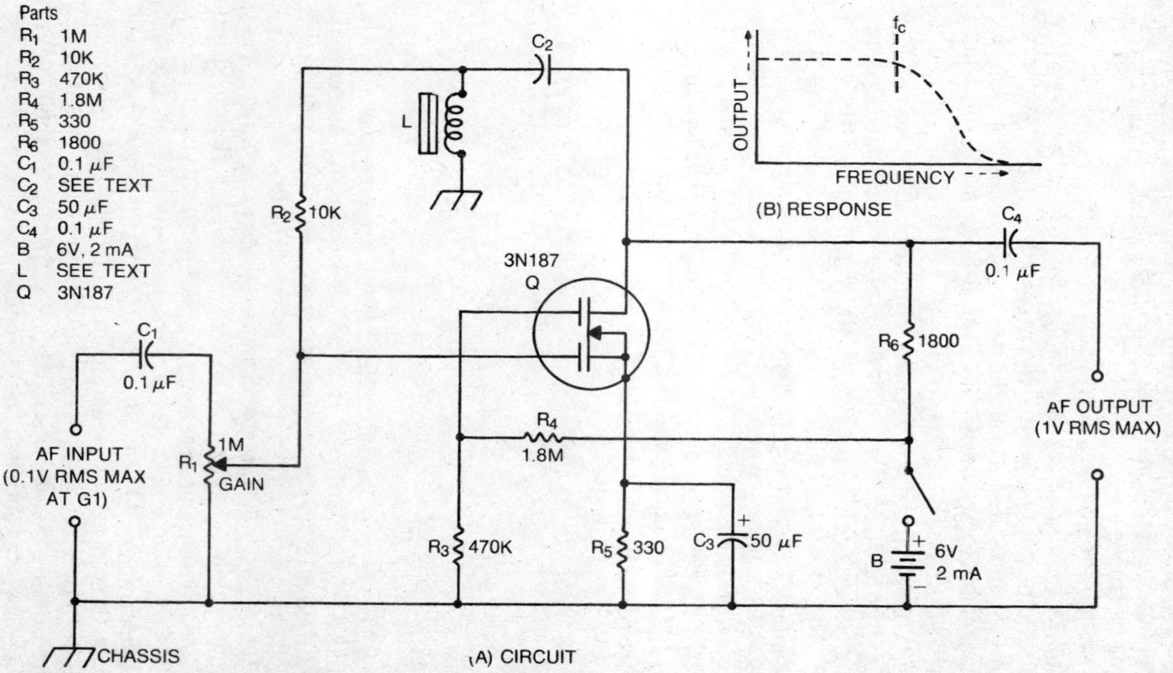

Fig. 2-13. LC-tuned low-pass amplifier.

RC-TUNED LOW-PASS AMPLIFIER

Figure 2-14A shows the circuit of an amplifier having the characteristics of a low-pass filter; but, unlike the preceding unit, this amplifier is tuned by means of a resistance–capacitance section (R_6–C_3), instead of an inductance–capacitance section. The advantage is compactness (inductors are bulky for AF tuning and are subject to pickup); however, the response shown in Fig. 2-14B is not so sharp as that of the *LC*-tuned amplifier (see, for comparison, Fig. 2-13B). In Fig. 2-14B, f_c is the cutoff frequency.

In this circuit, low-pass response is obtained by incorporating an *RC*-type *high*-pass filter section R_6–C_3 in the negative feedback loop between the drain output and gate-1 input circuit of the MOSFET, with the result that negative feedback occurs at all frequencies above f_c, canceling the gain at those frequencies, whereas the amplifier has approximately full gain below f_c.

Resistance R_6 and capacitance C_3 are chosen for the desired cutoff frequency according to the following equations, in which R_6 is in ohms, C_3 in farads, and f_c in hertz:

$$R_6 = \frac{1}{6.28 f_c C_3} \qquad \text{(Eq. 2-5)}$$

$$C_3 = \frac{1}{6.28 f_c R_6} \qquad \text{(Eq. 2-6)}$$

From Eq. 2-5, it is seen that for a cutoff frequency of 1000 Hz and a selected value of $C = 0.01\ \mu\text{F}$, $R_6 = 15{,}924\Omega$. In general, it will be better to select a capacitance, as above, and prune the resistance for the exact R_6 value from Eq. 2-5.

In the basic circuit, MOSFET *Q* receives its negative gate-1 bias from the voltage drop resulting from the flow of drain current through source resistor R_4, and its positive gate-2 bias from voltage divider R_2–R_3. With an individual MOSFET resistance R_3 may need some adjustment for linear response of the amplifier. Current drain from the 6V supply is approximately 2 mA. Open-circuit voltage gain of the circuit is approximately 10 in the passband. The maximum input signal voltage before output peak clipping is 0.1V RMS (higher amplitude signals are reduced to this maximum by appropriate settings of potentiometer R_1). The corresponding maximum output signal voltage is 1V RMS.

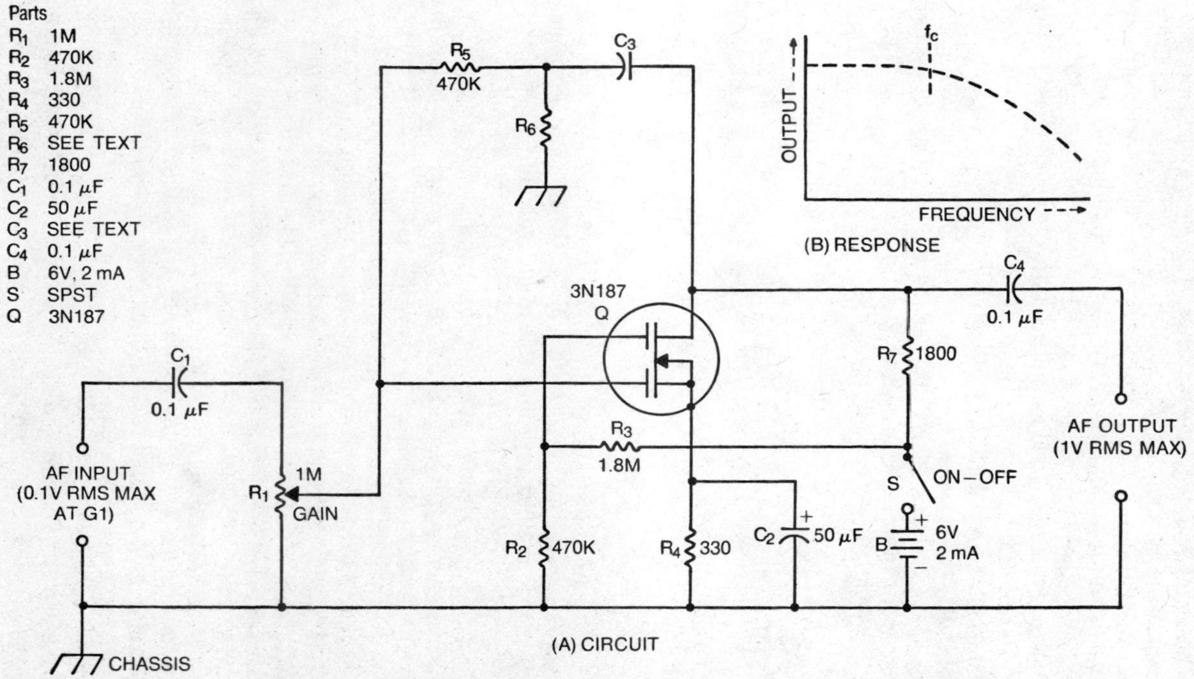

Fig. 2-14. RC-tuned low-pass amplifier.

LC-TUNED HIGH-PASS AMPLIFIER

Figure 2-15A shows the circuit of an amplifier having the characteristics of a high-pass filter. Figure 2-15B shows the amplifier response; here, f_c is the cutoff frequency. This response is obtained by incorporating an *LC*-type *low*-pass filter section $L-C_2$ in the negative feedback loop between the drain output and gate-1 input circuit of the MOSFET, with the result that negative feedback occurs at all frequencies below f_c, canceling the gain at those frequencies, whereas the amplifier has approximately full gain above f_c.

Inductance L and capacitance C_2 are chosen for the desired cutoff frequency f_c according to the following equations, in which L is in henrys, C in farads, and f_c in hertz:

$$L = \frac{10^4}{3.14 f_c} \qquad \text{(Eq. 2-7)}$$

$$C_2 = \frac{1}{3.14 f_c \times 10^4} \qquad \text{(Eq. 2-8)}$$

From Eq. 2-7, it is seen that for a cutoff frequency of 1000 Hz, $L = 3.18\text{H}$; and from Eq. 2-8, $C_2 = 0.032\ \mu\text{F}$. Inductor L should be of as high quality as practicable to insure that it has high Q; otherwise, the filter rolloff will not be sharp.

In the basic circuit, MOSFET Q receives its negative gate-1 bias from the voltage drop resulting from the flow of drain current through source resistor R_4, and its positive gate-2 bias from voltage divider R_2-R_3. With an individual MOSFET, resistance R_3 may need some adjustment for linear response of the amplifier. Current drain from the 6V supply is approximately 2 mA. Open-circuit voltage gain of the circuit is approximately 10 in the passband. The maximum input signal voltage before output peak clipping is 0.1V RMS at gate 1 (higher amplitude signals are reduced to this maximum by appropriate settings of potentiometer R_1). The corresponding maximum output signal voltage is 1V RMS.

RC-TUNED HIGH-PASS AMPLIFIER

Figure 2-16A shows the circuit of an amplifier having the characteristics of a high-pass filter; but, unlike the unit described in the preceding section, this amplifier is tuned by means of a resistance–capacitance section (R_3-C_2), instead of an inductance–capacitance section. The advantage is

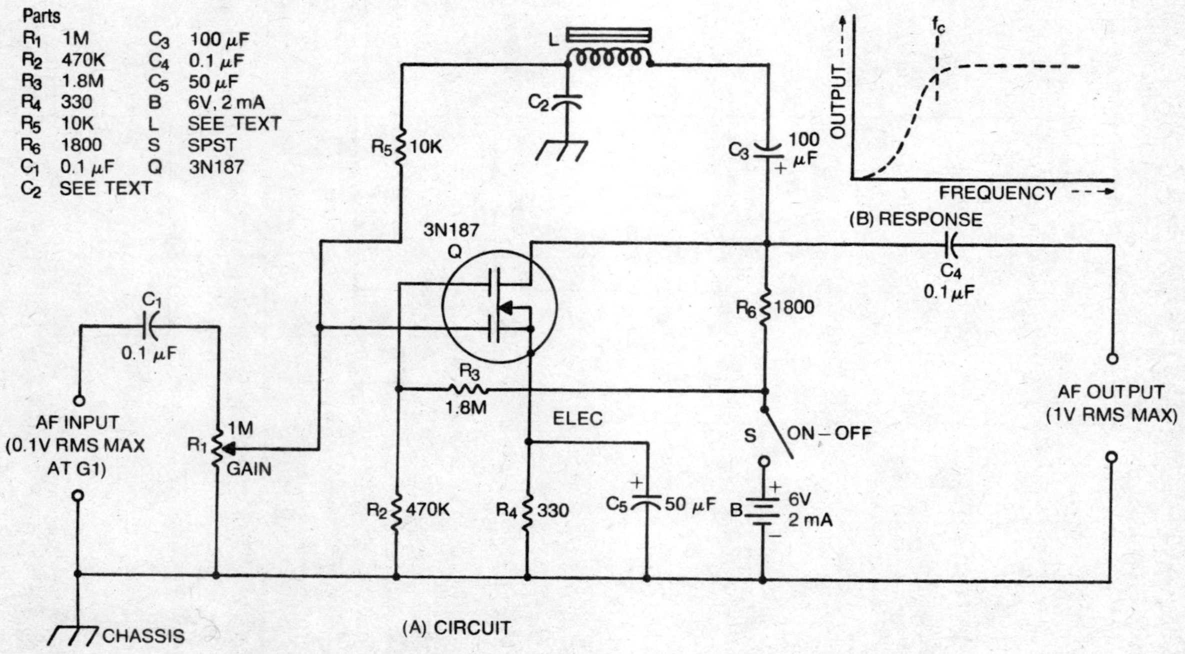

Fig. 2-15. LC-tuned high-pass amplifier.

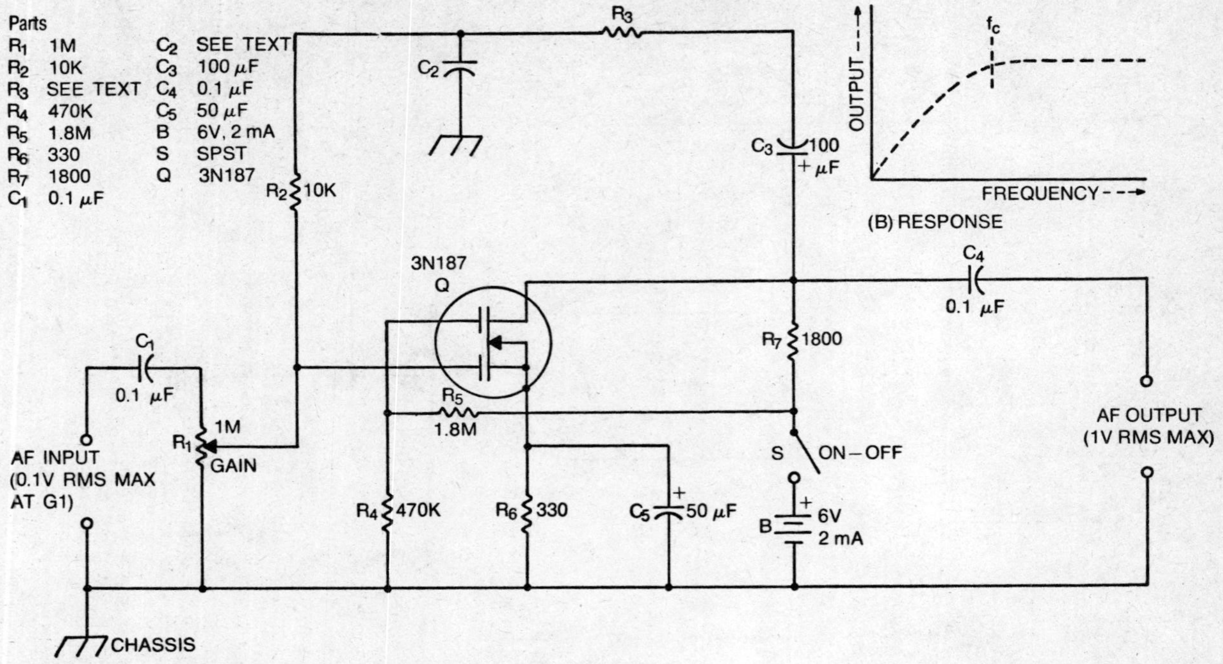

Fig. 2-16. RC-tuned high-pass amplifier.

compactness (inductors for AF tuning are bulky and subject to pickup); however, the response shown in Fig. 2-16B is not so sharp as that of the *LC*-tuned amplifier (see, for comparison, Fig. 2-15B). In Fig. 2-16B, f_c is the cutoff frequency.

In this circuit, high-pass response is obtained by incorporating an *RC*-type *low*-pass filter section (R_3–C_2) in the negative feedback loop between the drain output and gate-1 input circuit of the MOSFET, with the result that negative feedback occurs at all frequencies below f_c, canceling the gain at those frequencies, whereas the amplifier has approximately full gain above f_c.

Resistance R_3 and capacitance C_2 are chosen for the desired cutoff frequency according to the following equations, in which R_3 is in ohms, C_2 in farads, and f_c in hertz:

$$R_3 = \frac{1}{6.28 f_c C_2} \qquad \text{(Eq. 2-9)}$$

$$C_2 = \frac{1}{6.28 f_c R_3} \qquad \text{(Eq. 2-10)}$$

From Eq. 2-9, it is seen that for a cutoff frequency of 1000 Hz and a selected capacitance C_2 of 0.1 μF, resistance R_3 equals 1592Ω. In general, it will be better to select a capacitance, as above, and prune the resistance for the exact required R_3 value from Eq. 2-9.

In the basic circuit, MOSFET *Q* receives its negative gate-1 bias from the voltage drop resulting from the flow of drain current through source resistor R_6, and its positive gate-2 bias from voltage divider R_4–R_5. With an individual MOSFET, resistance R_5 may need some adjustment for linear response of the amplifier. Current drain from the 6V supply *B* is approximately 2 mA. Open-circuit voltage gain of the circuit is approximately 10 in the passband. The maximum input signal voltage before output peak clipping is 0.1V RMS (higher amplitude signals are reduced to this maximum by appropriate settings of potentiometer R_1). The corresponding maximum output signal voltage is 1V RMS.

LC-TUNED PEAK AMPLIFIER

Figure 2-17A shows the circuit of a sharply tuned peak amplifier that tends to pass a single frequency (signal peak). Such an amplifier is useful in null detection, signal separation,

Parts

R_1	1M	C_3	SEE TEXT
R_2	10K	C_4	100 μF
R_3	470K	C_5	0.1 μF
R_4	1.8M	B	6V, 2 mA
R_5	330	S	SPST
R_6	1800	L	SEE TEXT
C_1	0.1 μF	Q	3N187
C_2	50 μF		

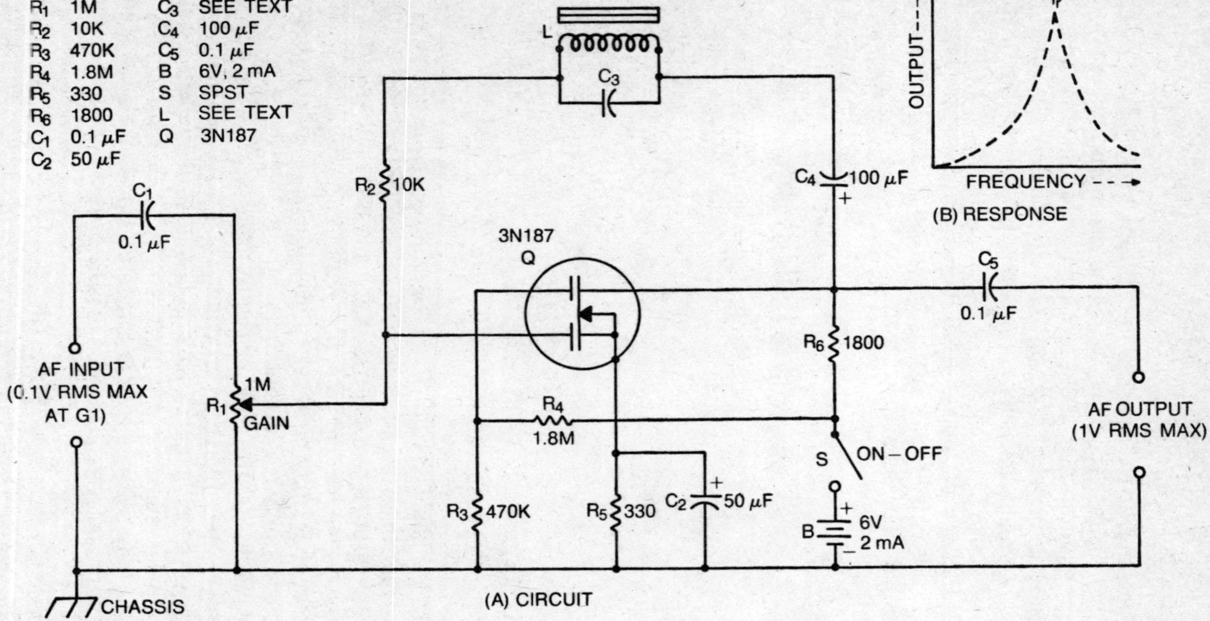

Fig. 2-17. LC-tuned peak amplifier.

distortion measurements, and similar applications in which it is desired to pass a single frequency while eliminating all others. In this circuit, frequency peaking is obtained by incorporating a signal *absorption* circuit (parallel-resonant circuit $L-C_3$) in the negative feedback loop between the drain output and gate-1 input section of the MOSFET, with the result that negative feedback occurs at all frequencies above and below the resonant frequency (f_r in Fig. 2-17B), which is absorbed by the $L-C_3$ combination, canceling the gain at those frequencies, whereas the amplifier has approximately full gain at frequency f_r. Figure 2-17B shows this response.

Inductance L and capacitance C_3 are chosen for the desired peak frequency according to the following equations, in which L is in henrys, C in farads, and f_r in hertz:

$$L = \frac{1}{39.5 f_r^2 C_3} \qquad \text{(Eq. 2-11)}$$

$$C_3 = \frac{1}{39.5 f_r^2 L} \qquad \text{(Eq. 2-12)}$$

From Eq. 2-12, it is seen that for a resonant (peak) frequency of 1000 Hz and a selected inductance of 10H, capacitance C_3 equals 0.0025 μF. In general, it will be better to select an inductance, as above, and prune the capacitance for the exact required C_3 value from Eq. 2-12.

In the basic circuit, MOSFET Q receives its negative gate-1 bias from the voltage drop resulting from the flow of drain current through source resistor R_5, and its positive gate-2 bias from voltage divider R_3-R_4. With an individual MOSFET, resistance R_4 may need some adjustment for linear response of the amplifier. Current drain from the 6V supply is approximately 2 mA. Open-circuit voltage gain of the circuit is approximately 10 at the peak (f_r in Fig. 2-17B). The maximum input signal voltage before output peak clipping is 0.1V RMS (higher amplitude signals are reduced to this maximum by appropriate settings of potentiometer R_1). The corresponding maximum output signal voltage is 1V RMS.

RC-TUNED PEAK AMPLIFIER

Figure 2-18A shows the circuit of a sharply tuned peak amplifier having characteristics similar to the one just described; but, unlike that unit, this amplifier is tuned by means of a resistance–capacitance section

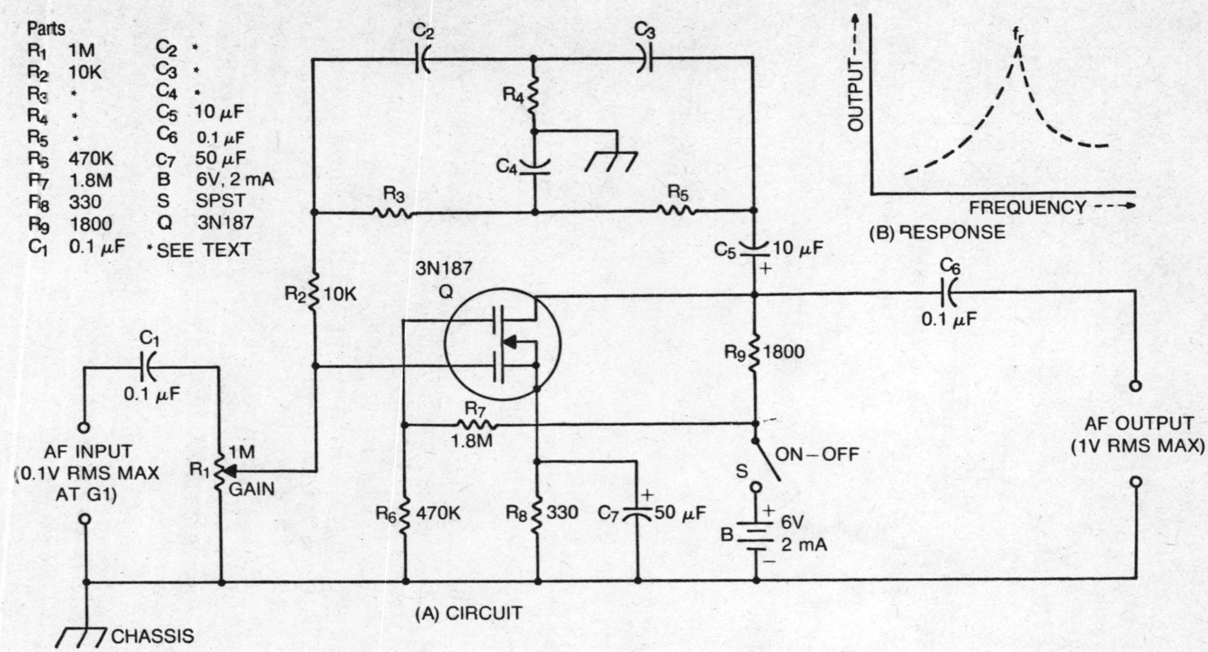

Fig. 2-18. RC-tuned peak amplifier.

C_2–C_3–C_4–R_3–R_4–R_5, instead of an inductance–capacitance section. The advantage is compactness (AF inductors are bulky and are subject to pickup). The response of the amplifier is shown in Fig. 2-18B, where f_r is the peak frequency.

In this circuit, peaking is obtained by incorporating a *null* circuit in the negative feedback loop between the drain output and gate-1 input section of the MOSFET, with the result that negative feedback occurs at all frequencies except the null frequency of the *RC* network (f_r in Fig. 2-18B), which is removed by the *RC* network, canceling the gain at those frequencies, whereas the amplifier has full gain at frequency f_r. Figure 2-18B shows this response.

The null circuit, which tunes the amplifier, is a parallel-*T* network (C_2–C_3–C_4–R_3–R_4–R_5). In this network, $C_2 = C_3 = 0.5C_4$, and $R_3 = R_5 = 2R_4$. When these relationships are preserved, the null frequency of the network, and therefore the peak frequency of the amplifier, may be determined by the equation

$$f_r = \frac{1}{6.28R_3C_2} \qquad \text{(Eq. 2-13)}$$

where f_r is in hertz, R_3 in ohms, and C_2 in farads.

From Eq. 2-13, it is seen that for a parallel-*T* circuit in which C_2 and C_3 are 0.1 μF, $C_4 = 2C_3 = 0.2$ μF, R_3 and R_5 are each 1592Ω, and $R_4 = 0.5R_3 = 796\Omega$, and the null frequency is 1000 Hz (the peak frequency f_r of the amplifier). In general, it is best to select the three capacitances ($C_2 = C_3 = 0.5C_4$) and then to adjust the resistances to exact values ($R_3 = R_5 = 2R_4$) as required. The following equations—in which R is in ohms, C in farads, and f_r in hertz—will help:

$$R_3 = R_5 = \frac{1}{6.28f_rC_2} \qquad \text{(Eq. 2-14)}$$

$$R_4 = 0.5R_3 \text{ (from Eq. 2-14)} \qquad \text{(Eq. 2-15)}$$

Resistances and capacitances must have exact values. The sharpness of the selectivity curve in Fig. 2-18B will depend upon the closeness with which these components meet specified values. The circuit may be made continuously tunable by using a 3-gang potentiometer for R_3–R_4–R_5 and switching the capacitors in groups of three to change frequency ranges. In this way, the amplifier can be given tuning ranges of 20–200, 200–2000, and 2000–20,000 Hz, to cover the audio-frequency spectrum.

In this circuit, MOSFET Q receives its negative gate-1 bias from the voltage drop resulting from the flow of drain current through source resistor R_8, and its positive gate-2 bias from voltage divider R_6–R_7. With an individual MOSFET, resistance R_7 may need some adjustment for linear response of the amplifier. Current drain from the 6V supply is approximately 2 mA. Open-circuit voltage gain is approximately 10 at the peak (f_r in Fig. 2-18B). The maximum input signal voltage before output peak clipping is 0.1V RMS (higher amplitude signals are reduced to this maximum by appropriate settings of potentiometer R_1). The corresponding maximum output signal voltage is 1V RMS.

LC-TUNED NOTCH AMPLIFIER

Figure 2-19A shows the circuit of a sharply tuned bandstop amplifier that tends to suppress a single frequency, i.e., to produce a *notch* or *slot* in the amplifier frequency response. Such an amplifier has applications in signal removal, distortion measurement, and similar uses where it is desired to remove a single frequency while passing all others. In this circuit, signal removal is accomplished by incorporating a parallel-resonant (wavetrap) circuit L–C_5 between the two stages of the MOSFET amplifier.

Inductance L and capacitance C_5 are chosen for the desired slot frequency (f_r in Fig. 2-19B) according to the following equations, in which L is in henrys, C in farads, and f_r in hertz:

$$L = \frac{1}{39.5 f_r C_5} \qquad \text{(Eq. 2-16)}$$

$$C_5 = \frac{1}{39.5 f_r^2 L} \qquad \text{(Eq. 2-17)}$$

In most instances, the inductor will have a fixed value. This means that L thus must be chosen and capacitance C_5 determined in terms of L. For example, if we start with an available 2H inductor and desire a slot frequency of 1000 Hz, capacitance C_5 (from Eq. 2-17) will be 0.0126 μF. To determine the frequency of any available inductor and capacitor, use the following equation, in which C_5 is in farads, L in henrys, and f_r in hertz:

$$f_r = \frac{1}{6.28\sqrt{LC_5}} \qquad \text{(Eq. 2-18)}$$

Parts

R_1	1M	R_9	330	C_7	0.1 μF		
R_2	470K	R_{10}	1800	B	6V, 4 mA		
R_3	1.8M	C_1	0.1 μF	S	SPST		
R_4	330	C_2	50 μF	S.B.	0.1 μF		
R_5	1800	C_3	10 μF	L	SEE TEXT		
R_6	470K	C_4	1 μF	Q_1	3N187		
R_7	470K	C_5	SEE TEXT	Q_2	3N187		
R_8	1.8M	C_6	50 μF				

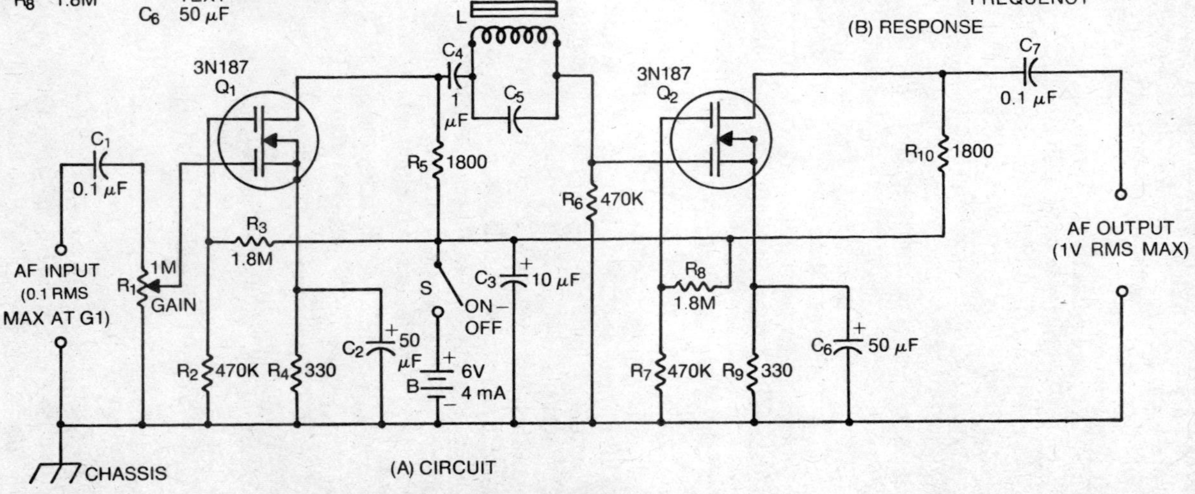

(A) CIRCUIT

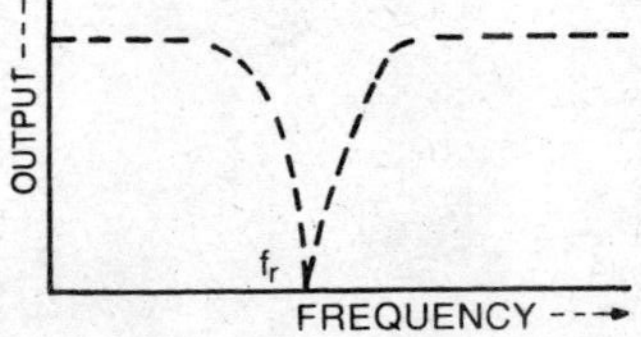

(B) RESPONSE

Fig. 2-19. LC-tuned notch amplifier

Thus, for an inductance of 10H and capacitance C_5 of 0.002 μF, $f_r = 1126$ Hz. For maximum sharpness of the response curve (Fig. 2-19B), the Q of inductor L must be high.

In this circuit, MOSFET Q_1 receives its negative gate-1 bias from the voltage drop resulting from the drain current through source resistor R_4, and its positive gate-2 bias from voltage divider $R_2 - R_3$. Similarly, MOSFET Q_2 receives its negative gate-1 bias from the voltage drop across source resistor R_9, and its positive gate-2 bias from voltage divider $R_7 - R_8$. With individual MOSFETs, resistors R_3 and R_8 may need some adjustment for linear operation of the amplifier. Current drain from the 6V supply is approximately 4 mA. Open-circuit voltage gain is approximately 90 to 100 in the passband, its exact value depending upon losses in inductor L. The maximum input signal voltage before output peak clipping is 10 mV RMS (higher amplitude signals are reduced to this maximum by appropriate settings of potentiometer R_1). The corresponding maximum output signal voltage is 1V RMS.

RC-TUNED NOTCH AMPLIFIER

Figure 2-20A shows the circuit of a sharply tuned notch amplifier having characteristics similar to those of the amplifier just described; but, unlike the unit just described, this amplifier is tuned by means of a resistance–capacitance null network ($C_4 - C_5 - C_6 - R_6 - R_7 - R_8$), instead of an inductance–capacitance section. The advantage is compactness, freedom from pickup, and adaptability to continuous tuning by varying the resistors. The response of the amplifier is shown by Fig. 2-20B, where f_r is the notch frequency (also called *slot* frequency).

The null circuit, which tunes the amplifier, is a parallel-T network ($C_4 - C_5 - C_6 - R_6 - R_7 - R_8$). In this network, $C_4 = C_5 = 0.5C_6$, and $R_6 = R_7 = 2R_8$. When these relationships are preserved, the null frequency of the network, and therefore the notch frequency in Fig. 2-20B, may be determined with the aid of the following equation (in which C_4 is in farads, R_6 in ohms, and f_r in hertz):

$$f_r = \frac{1}{6.28 R_6 C_4} \qquad \text{(Eq. 2-19)}$$

From Eq. 2-19, it is seen that for a parallel-T circuit in which C_4 and C_5 are each 0.005 μF, $C_6 = 0.01$ μF, R_6 and R_7 are each 1000Ω, and $R_8 = 500\Omega$, the null frequency is 31.85 kHz. In general, it is best to select the three capacitances ($C_4 = C_5 = 0.5C_6$) and make up the exact values of resistance ($R_6 = R_7 = 2R_8$), as required. The following equations—in

Parts

R_1	1M	R_{13}	1800	S	SPST
R_2	470K	C_1	0.1 μF	Q_1	3N187
R_3	1.8M	C_2	50 μF	Q_2	3N187
R_4	330	C_3	10 μF	*SEE TEXT	
R_5	1800	C_4	*		
R_6	*	C_5	*		
R_7	*	C_6	*		
R_8	*	C_7	50 μF		
R_9	10M	C_8	0.1 μF		
R_{10}	470K	C_9	10 μF		
R_{11}	1.8M	B	6V, 4 mA		
R_{12}	330				

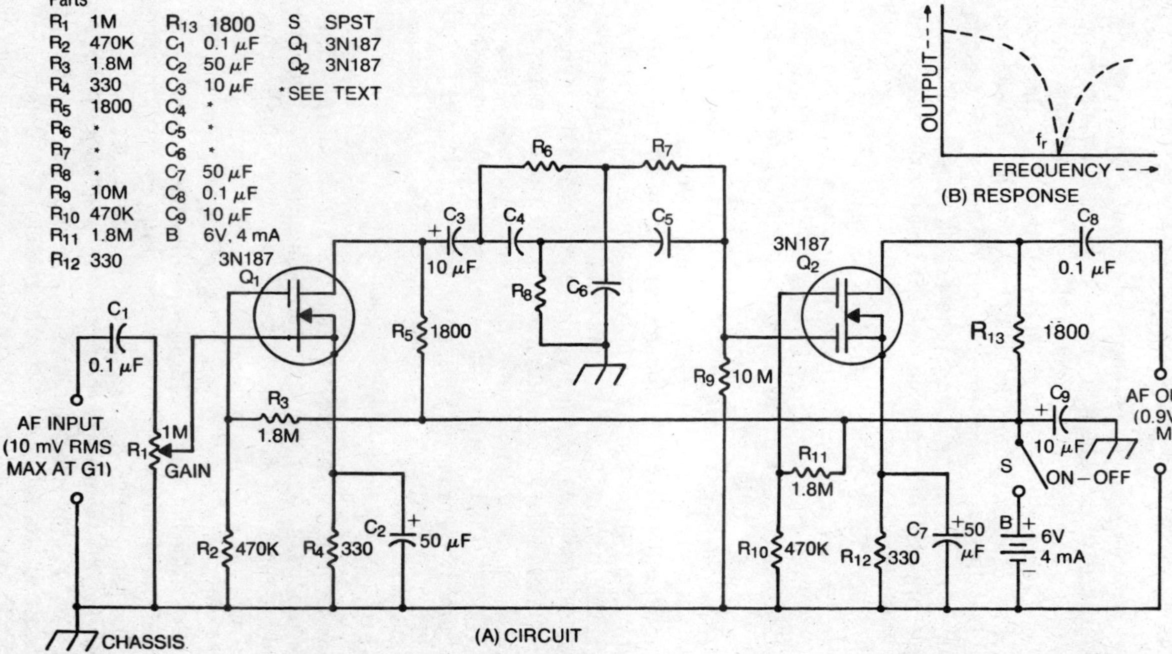

Fig. 2-20. RC-tuned notch amplifier.

which R is in ohms, C in farads, and f_r in hertz—will help:

$$R_6 = R_7 = \frac{1}{6.28 f_r C_4} \qquad \text{(Eq. 2-20)}$$

$$R_8 = 0.5R_6 \text{ (from Eq. 2-20)} \qquad \text{(Eq. 2-21)}$$

Resistances and capacitances must have exact values. The sharpness of the selectivity curve in Fig. 2-20B will depend upon the closeness with which these components meet specified values. The circuit may be made continuously tunable by using a 3-gang potentiometer for R_6–R_7–R_8 and switching the capacitors in groups of three to change frequency bands. In this way, the amplifier can be given tuning ranges, for example, of 20–200, 200–2000, and 2000–20,000 Hz.

In this circuit, MOSFET Q_1 receives its negative gate-1 bias from the voltage drop resulting from the flow of drain current through source resistor R_4, and its positive gate-2 bias from voltage divider R_2–R_3. Similarly, MOSFET Q_2 receives its negative gate-1 bias from the voltage drop across source resistor R_{12}, and its positive gate-2 bias from voltage divider R_{10}–R_{11}. With individual MOSFETs, resistors R_3 and R_{11} may need some adjustment for linear operation of the amplifier. Current drain from the 6V supply is approximately 4 mA. Open-circuit voltage gain is approximately 80 to 90 in the passband, depending upon individual MOSFETs and the amount of attenuation introduced by an individual parallel-*T* network. The maximum input signal voltage before output peak clipping is 10 mV RMS (higher amplitude signals are reduced to this maximum by appropriate settings of potentiometer R_1). The corresponding maximum output voltage is 0.8V to 0.9V RMS.

HEADPHONE AMPLIFIER

Although an amplifier for headphones is a relatively simple device, its operation also can be improved through the use of a MOSFET. The MOSFET affords very high input impedance and extreme miniaturization. The high impedance insures that the headphones, whether employed for communications or instrumentation, will present virtually no load to the signal source. This is especially important when magnetic headphones or a magnetic earpiece is used.

Figure 2-21 shows the circuit of a simple, inexpensive amplifier for high-resistance magnetic headphones. The input resistance of the circuit is approximately 1M, determined principally by the full resistance of volume control potentiometer R_1. A higher resistance may be employed, at some risk of stray pickup.

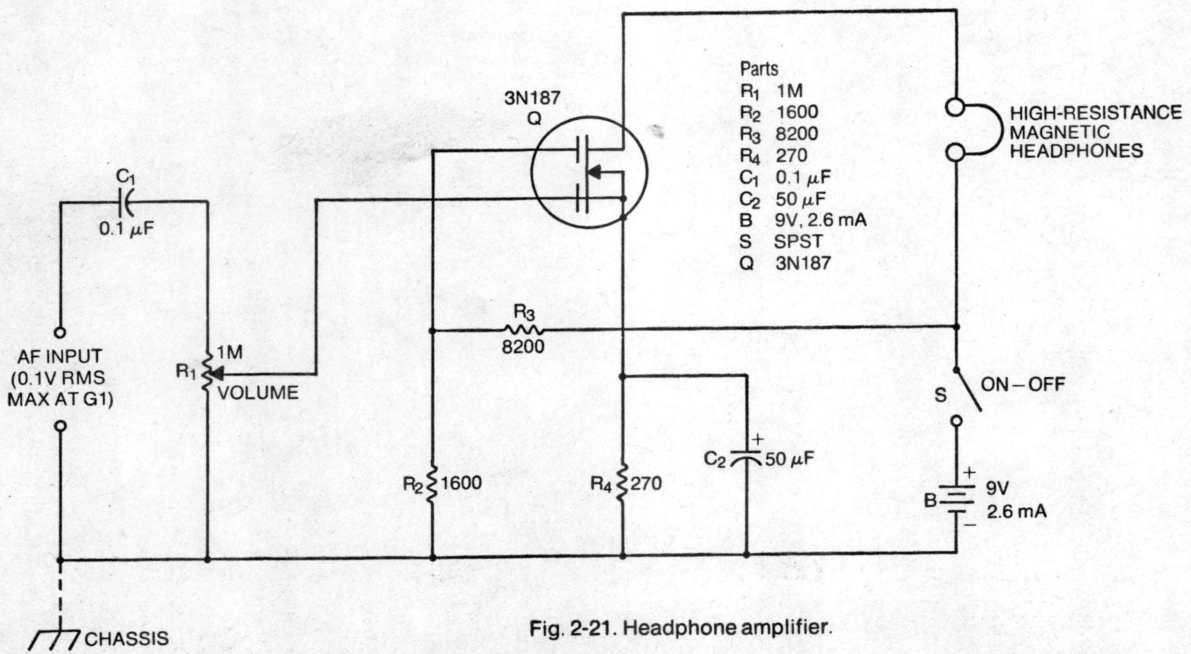

Fig. 2-21. Headphone amplifier.

In this circuit, the MOSFET receives its negative gate-1 bias from the voltage drop resulting from the flow of drain current through source resistor R_4, and its positive gate-2 bias from voltage divider R_2–R_3. With an individual MOSFET, resistance R_3 may need some adjustment for low distortion in the circuit. When a pair of 2000Ω magnetic headphones is used, the maximum input signal voltage before output peak clipping is 0.1V RMS (higher amplitude signals are reduced to this maximum by appropriate settings of potentiometer R_1). The corresponding maximum output voltage across the headphone load is 1V RMS. Under these conditions, the voltage gain is 10. A 0.5 mV RMS input signal at gate 1 of the 3N187 MOSFET gives a just barely audible signal in the headphones. Current drain from the 9V supply is approximately 2.6 mA.

PEAKED-RESPONSE HEADPHONE AMPLIFIER

Headphones usually have a resonant peak due chiefly to dimensions of the diaphragm. This peak may be intensified or a more satisfactory one provided simply by tuning the inductance of the headphones with a suitable capacitance such as C_3 in Fig. 2-22. Thus, the headphones may be tuned to a readily recognizable frequency such as 400 or 1000 Hz, and this will improve code reception, as well as providing additional signal selectivity. Such resonating of the headphones is advantageous also in the balancing of an AC bridge by means of headphones.

The tuning may be accomplished practically by feeding a signal of the desired frequency into the input terminals of the amplifier and adjusting a capacitor decade (connected in parallel with the headphones) for peak response, as indicated by maximum upswing of an AC electronic voltmeter (VTVM or TVM) connected temporarily across the headphones. Or the inductance L of the headphones may be measured with a suitable AC bridge and the required capacitance calculated with the aid of the following equation, in which L is the headphone inductance in henrys, C_3 is the required capacitance in microfarads, and f_r is the resonant frequency in hertz:

$$C_3 = \frac{1{,}000{,}000}{39.5 f_r^2 L} \qquad \text{(Eq. 2-22)}$$

In all respects other than the tuning of the headphones, this circuit is the same as the simpler one described in the preceding section. For electrical characteristics, therefore, refer to that section.

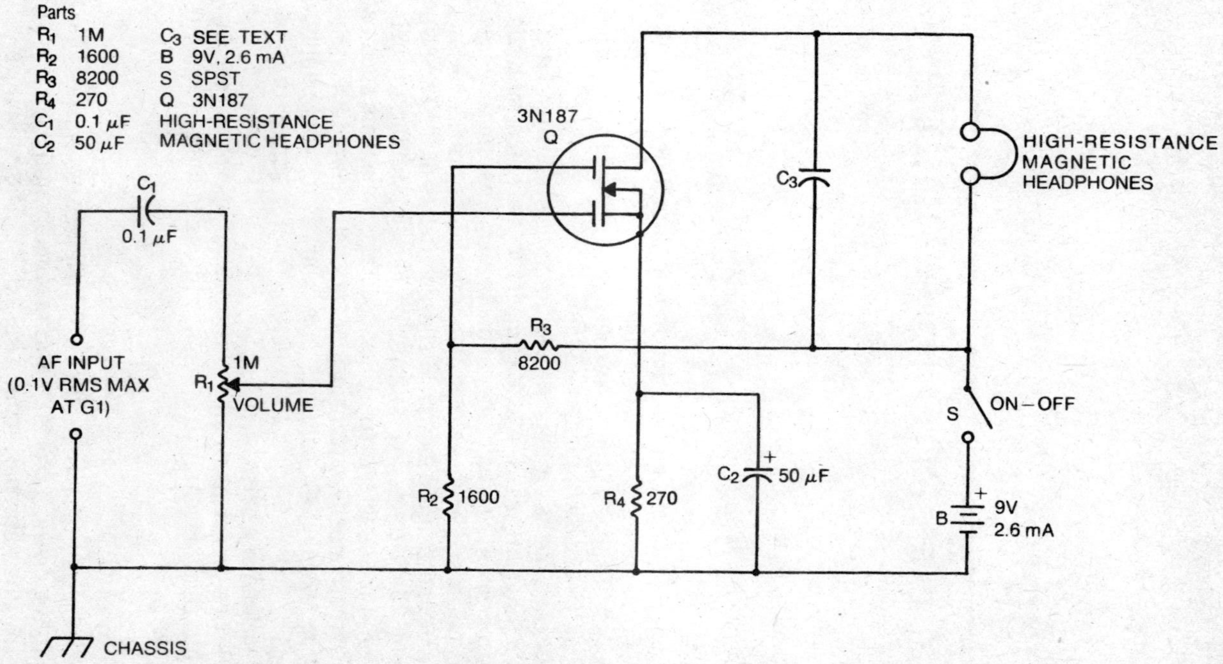

Fig. 2-22. Peaked-response headphone amplifier.

Chapter 3

DC, RF, and IF Amplifiers

This chapter offers circuits for four DC amplifiers, four RF amplifiers, one video amplifier, and three IF amplifiers. These circuits will suggest others to experimenters, and the intended applications also will suggest others.

In each of the circuits, unless otherwise indicated on the circuit diagram or in the text, capacitances are in picofarads, resistances in ohms, and inductances in microhenrys. Resistors are ½W, and electrolytic capacitors are 25 working volts. Because of the electrical characteristics of the MOSFET, most of the circuits are seen to be closely similar to their vacuum-tube counterparts. Thus, special components such as tapped transistor-type transformers are not required.

For simplicity, batteries are shown for DC supply; however, a *well filtered* line-operated power supply may be used instead of a battery.

Before undertaking the wiring and operation of any circuit, read carefully the hints and precautions given in Chapter 1.

DC AMPLIFIER

Figure 3-1 shows a simple single-MOSFET circuit for amplifying direct currents. In this arrangement the current of interest flows through input resistor R_1. The voltage drop which accordingly is developed across this resistor is applied to gate 1 of the 3N187 MOSFET. This causes the MOSFET drain current to increase in accordance with the transconductance

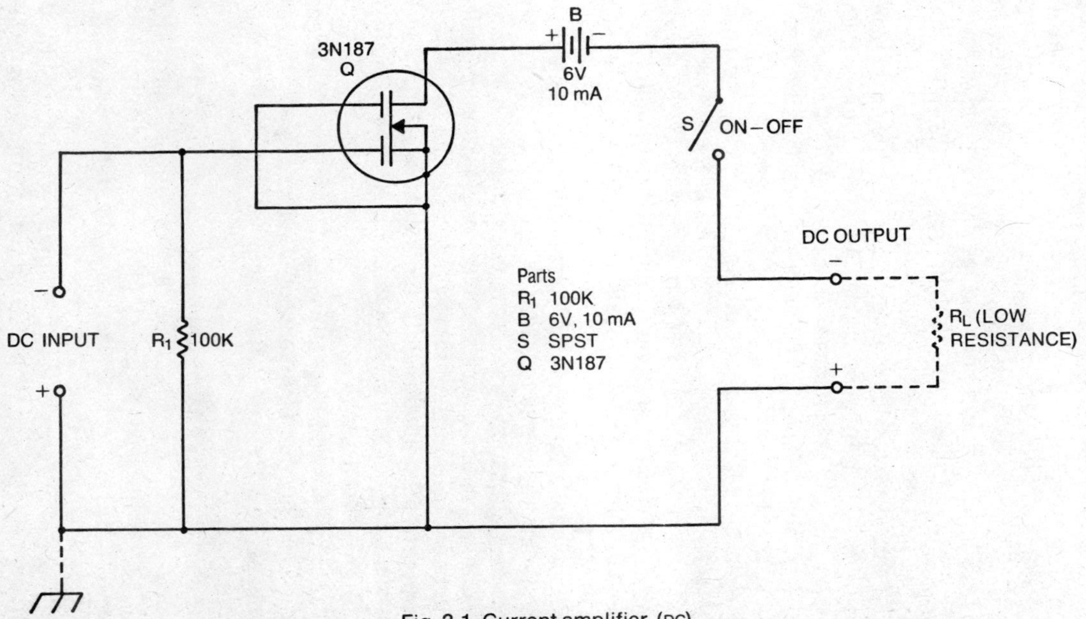

Fig. 3-1. Current amplifier (DC).

of the device, with the result that the output current, flowing through a low-resistance load, R_L, is much greater than the input current flowing through R_1. Gate 2 of the MOSFET is returned to the source.

A residual drain current of approximately 3 mA DC flows in the output circuit, and this quiescent current can be balanced out of the load device (such as a meter), when desired, by means of any one of the conventional DC-bucking circuits. When the input signal voltage drop applied to gate 1 is 1V, the drain current increases to approximately 6 mA. At 1V input, the current through resistor R_1 is 0.00001A = 0.01 mA, and for the corresponding drain current change of 3 mA, the resulting current gain is 300. A higher value of input resistance R_1 results in a lower input current for the 1V gate-1 signal and a correspondingly higher current gain. Thus, if R_1 = 1M, i_i = 1 μA, and the current gain is 3000.

The circuit has the disadvantage of high input resistance (high insertion resistance); however, this is of no consequence in some applications. A circuit oddity also is the floating battery *B*, but this too causes no trouble in most applications of a current amplifier such as this one.

Individual MOSFETS show different values of transconductance, so the circuit you put together can be more sensitive or less sensitive than the figures given here. Individual calibration therefore is required. The actual current amplification depends heavily on the value of the load resistance R_L, being highest for low R_L resistance.

In order to be adaptable to individual MOSFETS, the 6V supply should be capable of supplying 10 mA.

DC VOLTAGE AMPLIFIER

It is often desired to amplify a DC voltage without going into the complication of a chopper-stabilized type of circuit. For modest requirements, the circuit shown in Fig. 3-2 is recommended.

This arrangement is a simple source-follower circuit for DC signal input and output. It is the simplest DC voltage amplifier. With zero signal at the input terminals, the quiescent drain current produces a static voltage of 0.15V at the output terminals. (This is barely discernible on the scale of a 0–10 voltmeter connected to the output terminals; but if it is objectionable, it may be balanced out with the aid of any one of the conventional DC-bucking circuits.) When the input signal is 1V at gate 1 of the MOSFET (gain control R_1 adjusted for maximum gain), the corresponding voltage at the output terminals is 5.2V. The actual output voltage thus is 5.2–0.15 = 5.05V, and the voltage gain of the amplifier is 5.05.

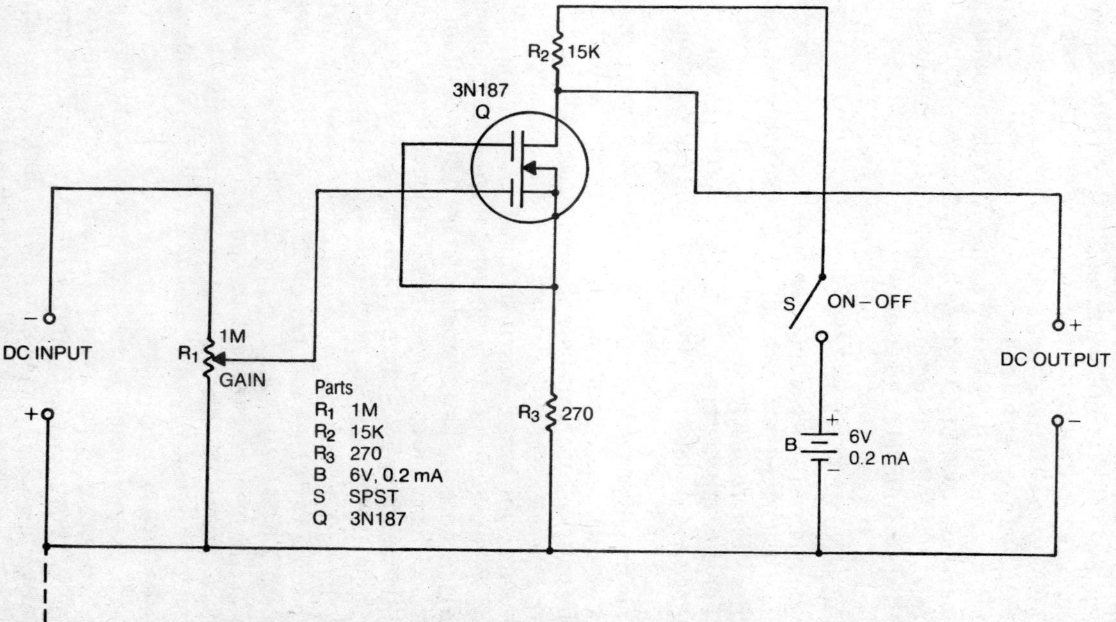

Fig. 3-2. Voltage amplifier (DC).

The maximum recommended DC input signal is 1V. (Higher signal voltages are reduced to this maximum by appropriate settings of potentiometer R_1). The corresponding maximum output signal voltage is 5.05V DC (after correction). For reduced loading of a DC signal source, the full resistance of potentiometer R_1 may be increased. Conversely, for reduced susceptibility to stray pickup, the maximum resistance of this potentiometer may be reduced.

Maximum drain from the 6V supply is approximately 0.2 mA. Individual MOSFETs will draw different operating current and will also offer different values of transconductance (affecting the gain figure given here).

DC GALVANOMETER AMPLIFIER

The input resistance of a DC galvanometer such as is used for bridge balancing may be greatly increased with a MOSFET amplifier. Figure 3-3 shows the circuit of a balanced DC amplifier for this purpose. This circuit provides an input resistance of 1M, determined largely by the full resistance of sensitivity control potentiometer R_1.

This amplifier itself consists of a 4-arm bridge circuit in which the two MOSFETs are in opposite arms. The bridge is initially balanced, in the manner of an electronic voltmeter, by means of the zero-set potentiometer R_5, and galvanometer M is connected to the bridge output terminals. A DC input signal applied to gate 1 of one of the MOSFETs (here, Q_1) unbalances the bridge and causes the meter to be deflected from its center-zero position of rest. A positive input signal at the amplifier input terminals drives the meter upscale; a negative input signal drives the meter downscale.

Meter M may be a 25—0—25 μA or 50—0—50 μA model. The internal resistance of a typical 25—0—25 μA instrument (Simpson Model 2123) is approximately 1800Ω. The amplifier provides 1M input resistance, thus apparently increasing the meter resistance 555 times. The increased resistance greatly enhances the sensitivity and resolution of the instrument in bridge balancing.

When potentiometer R_5 is adjusted, with zero input signal, galvanometer M is set exactly to center zero. Then with sensitivity control potentiometer R_1 set for maximum sensitivity, a +0.25V DC input signal will deflect meter M up to full scale, whereas a −0.25V will deflect the meter down to full scale.

Drain from the 6V supply is approximately 1.6 mA. Individual MOSFETs may exhibit somewhat different load current, but will not materially affect operation of the circuit.

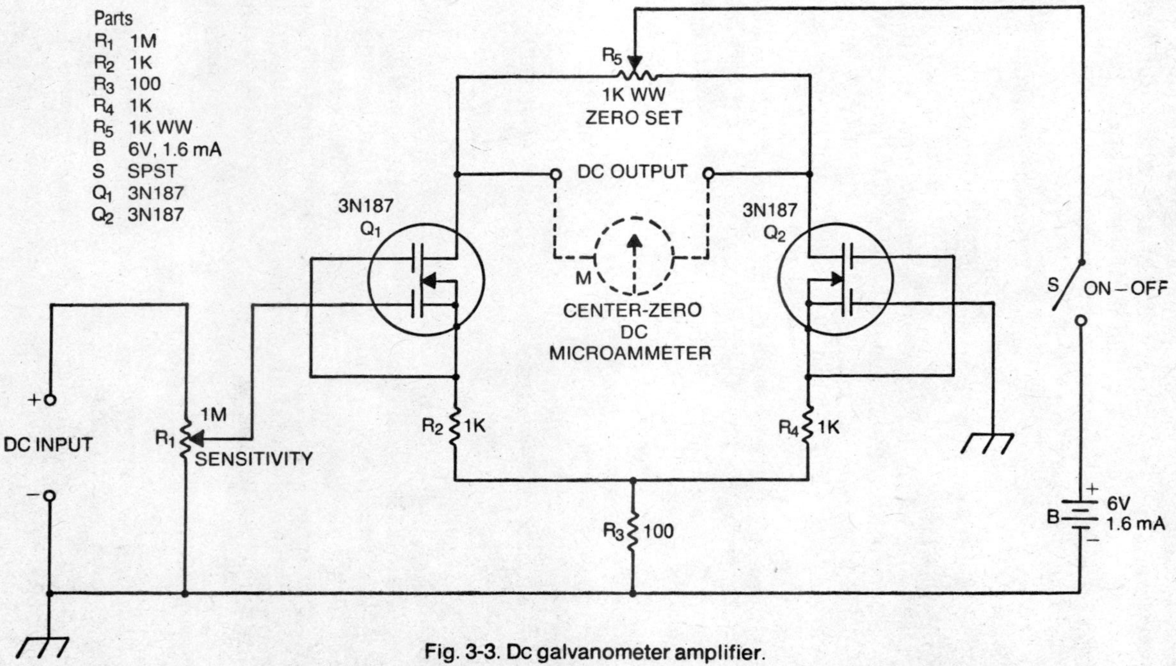

Fig. 3-3. Dc galvanometer amplifier.

DC SOURCE FOLLOWER

Figure 3-4 shows a source-follower circuit designed for direct-current use. Like the familiar AC source follower (and its counterparts, the emitter follower and cathode follower), this circuit provides "impedance" transformation; that is, it presents a high input resistance to a DC signal source, and a low output resistance to a load.

When the DC input signal e_i is zero, the quiescent drain current of the MOSFET, flowing through source resistor R_2, produces a static voltage e_0 at the output terminals of 0.75V. In many cases this initial voltage will be of no concern; in other instances, however, it may easily be balanced out with the aid of any one of the conventional bucking circuits. When the input signal is e_i is 1V, the output signal voltage is 1.65V. This is an output voltage change of 1.65−0.75 = 0.9V, and it represnets a voltage gain of 0.9 for the source follower.

The input resistance of the circuit is 470K, provided chiefly by the resistor R_1. The output resistance is 2000Ω, provided largely by resistor R_2. Other values of R_1 and R_2 may be employed, with somewhat different values of output resistance. If desired, R_1 may be a gain-control potentiometer. Response of the circuit is approximately linear.

Drain from the 6V source is approximately 1 mA. Individual MOSFETs will show different current drains, as well as different transconductance values (this latter affecting the source-follower voltage gain).

GENERAL-PURPOSE RF AMPLIFIER

Figure 3-5 shows the circuit of a single-stage outrigger RF amplifier which can provide substantial signal gain as a preselector or booster for receivers and instruments. The circuit is tuned with a dual 365 pF variable capacitor C_1−C_4. With standard plug-in coils, as shown in Fig. 3-5, the tuning range can be 500 kHz to 30 MHz. Table 3-1 gives winding

Table 3-1. Broadcast-Band Coil Data.

L_1	10 turns No. 32 enameled wire closewound around the lower end of L_2 and insulated from the latter.
L_2	130 turns No. 32 enameled wire closewound on a 1 in. diameter form.
L_3	130 turns No. 32 enameled wire closewound on a 1 in. diameter form.
L_4	12 turns No. 32 enameled wire closewound around the lower end of L_3 and insulated from the latter.

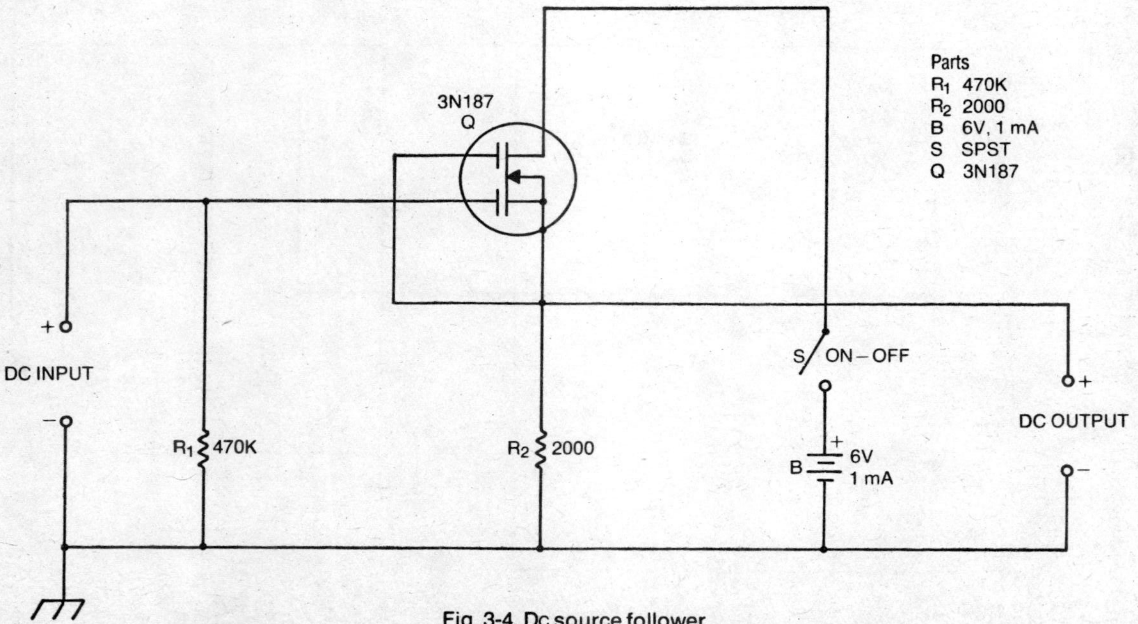

Fig. 3-4. Dc source follower.

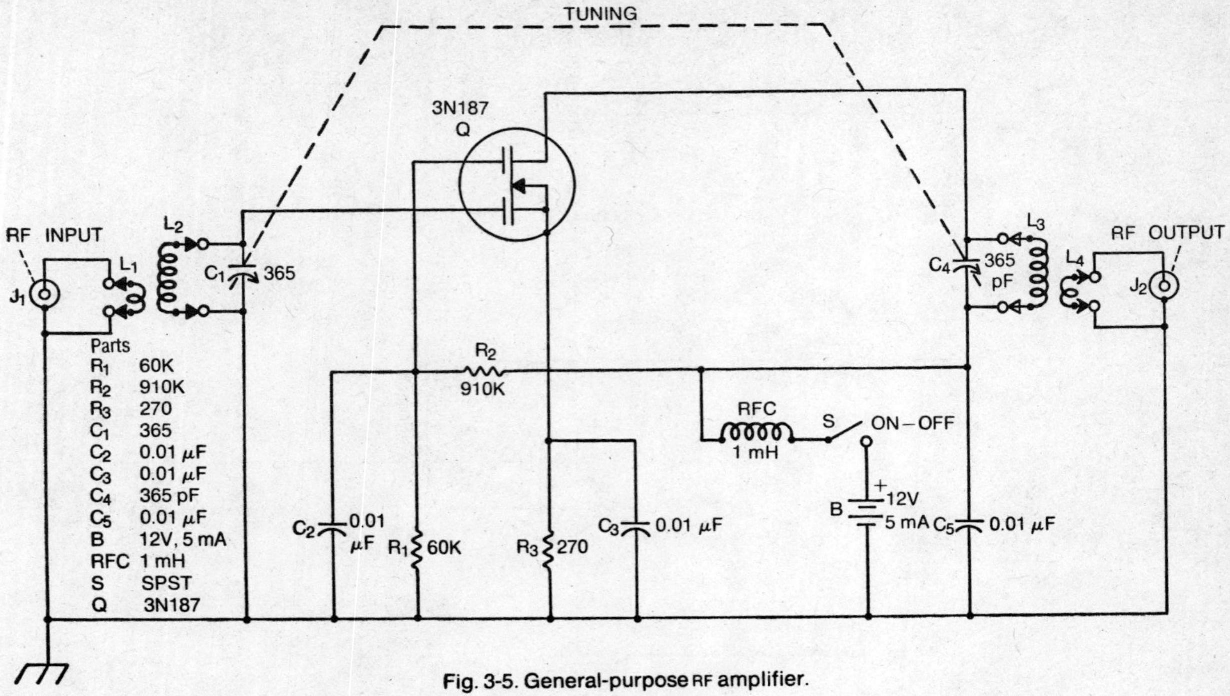

Fig. 3-5. General-purpose RF amplifier.

instructions for coils for the standard broadcast band. Owing to the extremely low value of interelectrode capacitance in the MOSFET, no neutralization is required.

In this circuit, the incoming RF signal, selected by the input tuned circuit, is presented to gate 1 of the MOSFET; the output signal is developed by drain current in the output tuned circuit. Coaxial jacks (J_1 and J_2) or other suitable terminals lead the signal into and out of the amplifier. The negative gate-1 bias is provided by the voltage drop resulting from the flow of drain current through source resistor R_3. Postive gate-2 bias is produced by voltage divider R_1-R_2. With an individual MOSFET, resistor R_2 may require some adjustment for maximum amplification. If the amplifier is to be operated above the standard broadcast band, shielding is recommended; i.e., the unit should be enclosed in an aluminum box.

Current drain from the 12V supply is approximately 5 mA. An individual MOSFET may show a somewhat different load current value.

SINGLE-ENDED TRANSMITTER-TYPE RF AMPLIFIER

The circuit given in Fig. 3-6 has an untuned input and a tuned output. This is the type of circuit found in the exciter and final-amplifier stages in transmitters, hence its designation *transmitter-type*. The extremely low interelectrode capacitance of the MOSFET enables this circuit to be operated without neutralization. The circuit may be employed as given in flea-powered transmitters and experimental equipment.

Tuning is accomplished with the single 100 pF variable capacitor C_5. Standard amateur-band plug-in coils may be used (each coil set contains an L_1 and L_2), or a single L_1-L_2 pair may be wound for a specific frequency, using coil-winding instructions found in the various radio handbooks. The circuit is entirely conventional and will be familiar to hams and experimenters, with the exception that MOSFET Q replaces the vacuum tube usually found in this circuit. Link-coupled. low-impedance output is provided by coupling coil L_2. In some installations, however, capacitive coupling will be desirable, and this can be provided by output capacitor C_6, which will be 100 to 250 pF, depending upon individual requirements. When capacitive output coupling is used, the small coil L_2 is omitted.

In this amplifier circuit, the MOSFET receives its negative gate-1 bias from the voltage drop resulting from the flow of drain current through source resistor R_4, and its positive gate-2 bias from voltage divider R_2-R_3. With an individual MOSFET, resistance R_3 may need some adjustment for

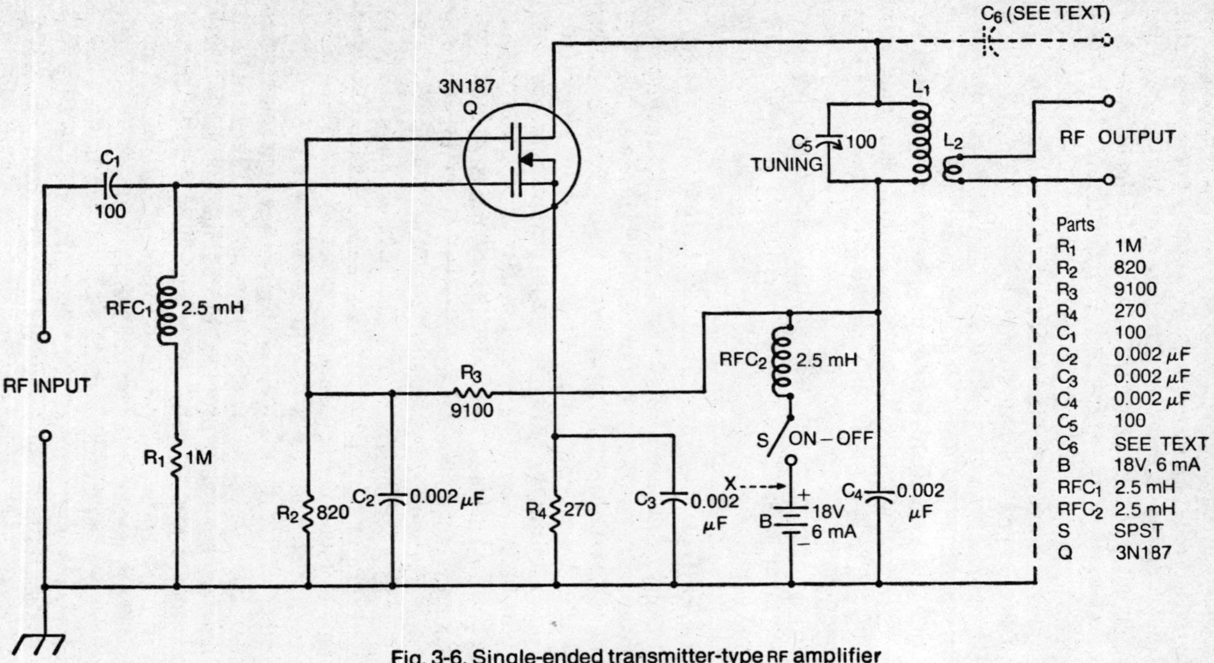

Fig. 3-6. Single-ended transmitter-type RF amplifier

maximum RF output. Maximum current drain from the 18V source is approximately 6 mA.

Tuning of the circuit is straightforward: (1) Connect a 0–10 mA DC meter temporarily in series with the battery and switch at point *X*. (2) Close switch *S*. (3) Disconnect any load from the output terminals. (4) Apply the input signal to the input terminals. (5) Adjust tuning capacitor C_5 for minimum dip of the milliammeter reading. (6) Connect the external load, making any adjustments in the coupling (coil L_2 or capacitor C_6) to bring the milliammeter reading up to 6 mA. (7) Retune C_5, if necessary, for resonance.

The RF power output of the amplifier is approximately 36 mW. For this output, the driving signal at the input terminals must be approximately 0.8V RMS. The circuit may be keyed for CW transmission or amplitude-modulated for voice transmission, in the usual manner.

PUSH-PULL TRANSMITTER-TYPE RF AMPLIFIER

The RF amplifier circuit shown in Fig. 3-7 is similar to the one just described and shown in Fig. 3-6, except that the Fig. 3-7 circuit employs two MOSFETs in push-pull for increased power output and good harmonic cancellation. The general description of the single-ended circuit (preceding section) applies also to this push-pull circuit, as do the tuning instructions.

In Fig. 3-7, MOSFET Q_1 receives its negative gate-1 bias from the voltage drop resulting from the flow of drain current through source resistor R_4, and its positive gate-2 bias from voltage divider R_2–R_3. Similarly, MOSFET Q_2 receives its negative gate-1 bias from the voltage drop resulting from the flow of drain current through source resistor R_7, and its positive gate-2 bias from voltage divider R_5–R_6. With individual MOSFETs, resistors R_3 and R_6 may require some adjustment for maximum RF power output.

The RF power output of the amplifier is approximately 72 mW. For this output, the driving signal (gate 1 to gate 1) must be approximately 1.6V RMS. Maximum current drain from the 18V source *B* is approximately 12 mA. The circuit may be keyed for CW transmission or amplitude-modulated for voice transmission, in the usual manner.

100 MHz RF AMPLIFIER

Figure 3-8 shows the circuit of a radio-frequency amplifier designed especially for operation at 100 MHz. The single 3N187 MOSFET is operated, together with its associated components, inside a metal shield box. The circuit, which operates between

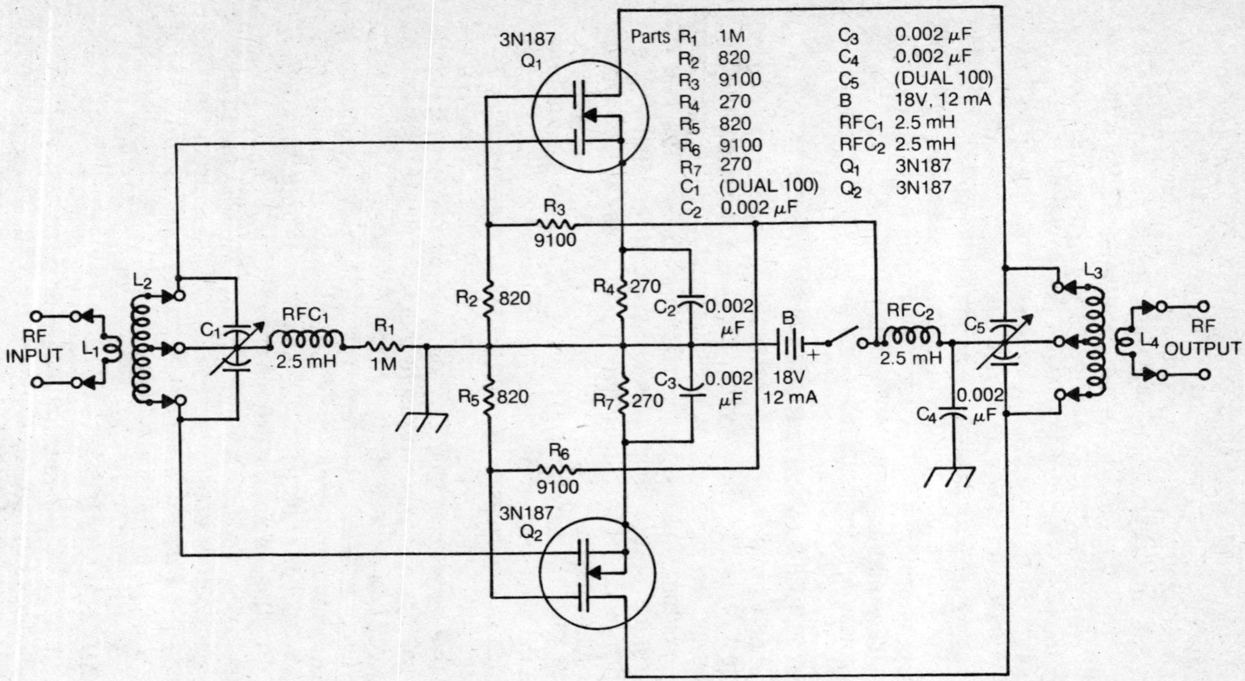

Fig. 3-7. Push-pull transmitter-type RF amplifier.

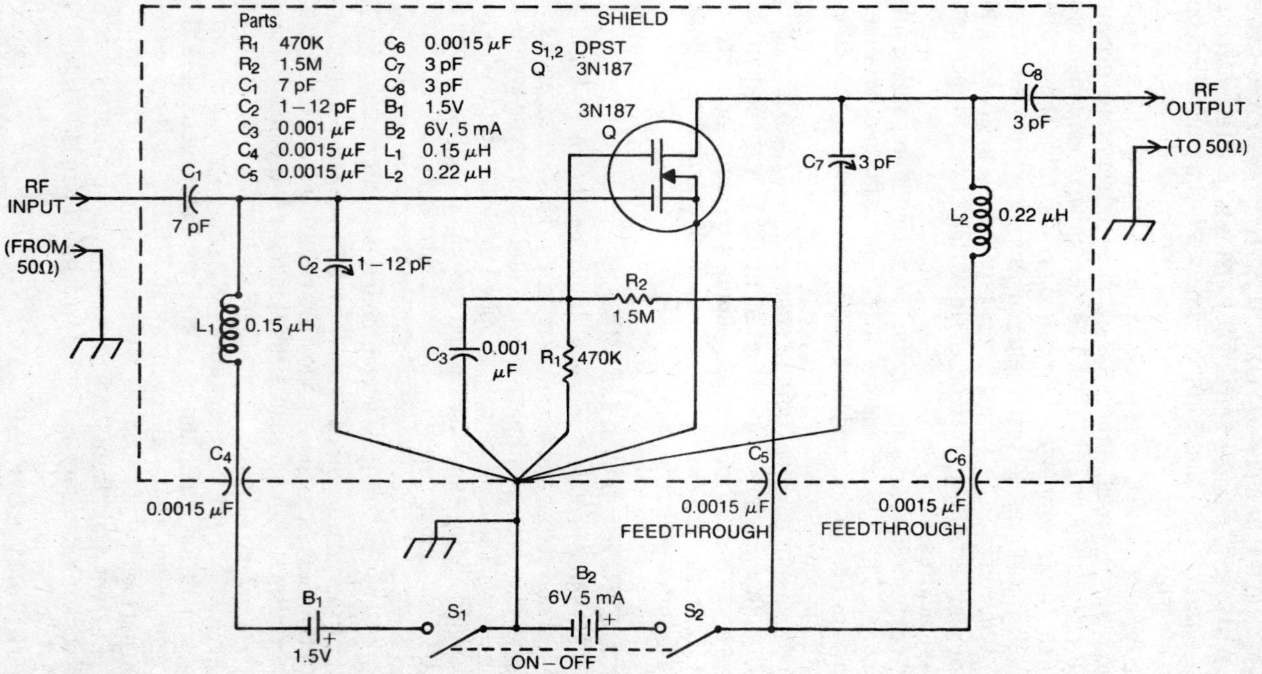

Fig. 3-8. 100 MHz RF amplifier.

a 50Ω signal source and 50Ω load, is tuned at both input (L_1–C_2) and output (L_2–C_7) ends. These two tuned circuits are adjusted for maximum RF output. Because the interelectrode capacitance of the MOSFET is extremely low, no neutralization is required.

Various portions of the circuit are grounded to a single point on the shield (chassis), as shown; separate ground returns must not be employed. Feedthrough bypass capacitors (C_4, C_5, C_6) are employed for strategic leads. All batteries and switches are outside the shield. In this amplifier, all leads must be run in as short and direct a path as practicable, and preferably should be straight. Rigid wiring is very important, as is lead dress.

Commercial parts are usable; no special components are required. The 0.15 μH input coil (L_1), for example, is Miller 4582-E or equivalent; and the 0.22 μH output coil (L_2) is Miller 4584-E or equivalent.

Maximum current from 6V source B_2 is approximately 5 mA, although this can vary somewhat with individual MOSFETs. The 1.5V battery, B_1, supplies virtually no current. Both batteries are switched simultaneously by the double-pole, single-throw switch S_1–S_2.

SINGLE-TUNED IF AMPLIFIER

The intermediate-frequency amplifier shown in Fig. 3-9 has a tuned output and an untuned output. This arrangement is desirable in certain simple receivers and heterodyne-type test equipment. The input signal is applied to gate 1 of the MOSFET through coupling capacitor C_1, and the output signal is coupled through a standard, tube-type IF transformer T for the desired intermediate frequency. Either low-frequency or high-frequency IF may be accommodated by this circuit.

In this arrangement, gate 1 of the MOSFET receives its negative bias from the voltage drop resulting from the flow of drain current through source resistor R_4, and the positive bias for gate 2 is supplied by voltage divider R_2–R_3. With an individual MOSFET, resistance R_3 may require some adjustment for maximum IF gain. Extensive shielding of the circuit is not required, even at 10.7 MHz, so long as IF transformer T is shielded.

Current drain from the 9V source is approximately 3 mA, although this can vary somewhat with individual MOSFETs.

DOUBLE-TUNED IF AMPLIFIER

Figure 3-10 shows the circuit of a double-tuned intermediate-frequency amplifier. This circuit may be used at

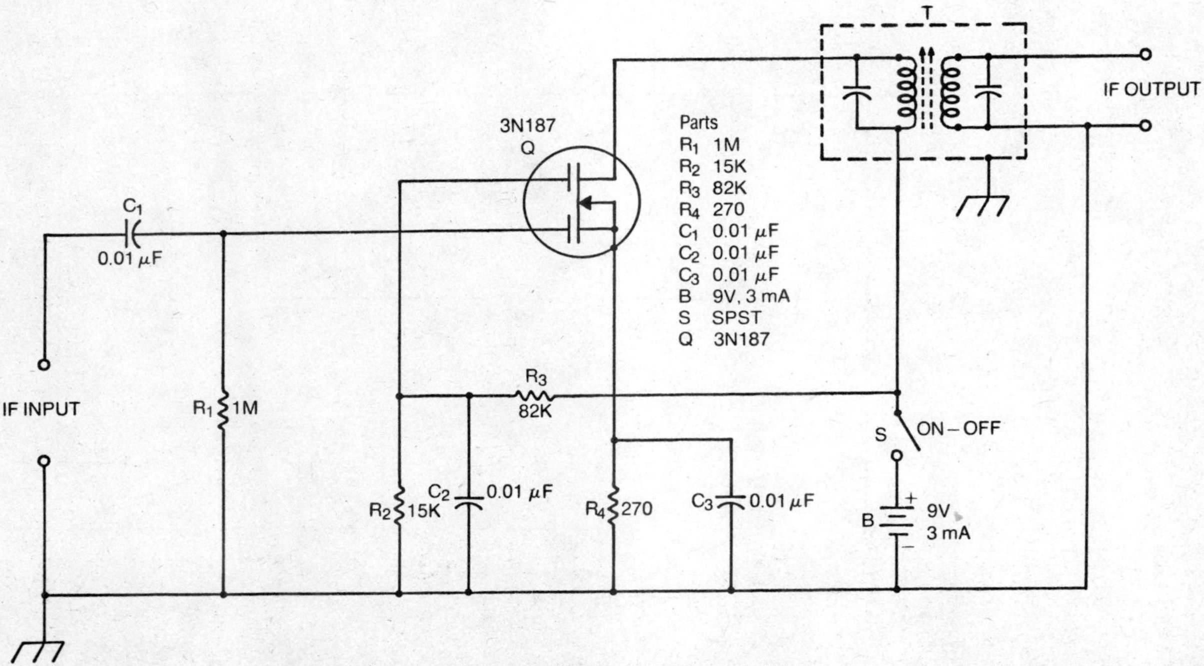

Fig. 3-9. Single-tuned IF amplifier.

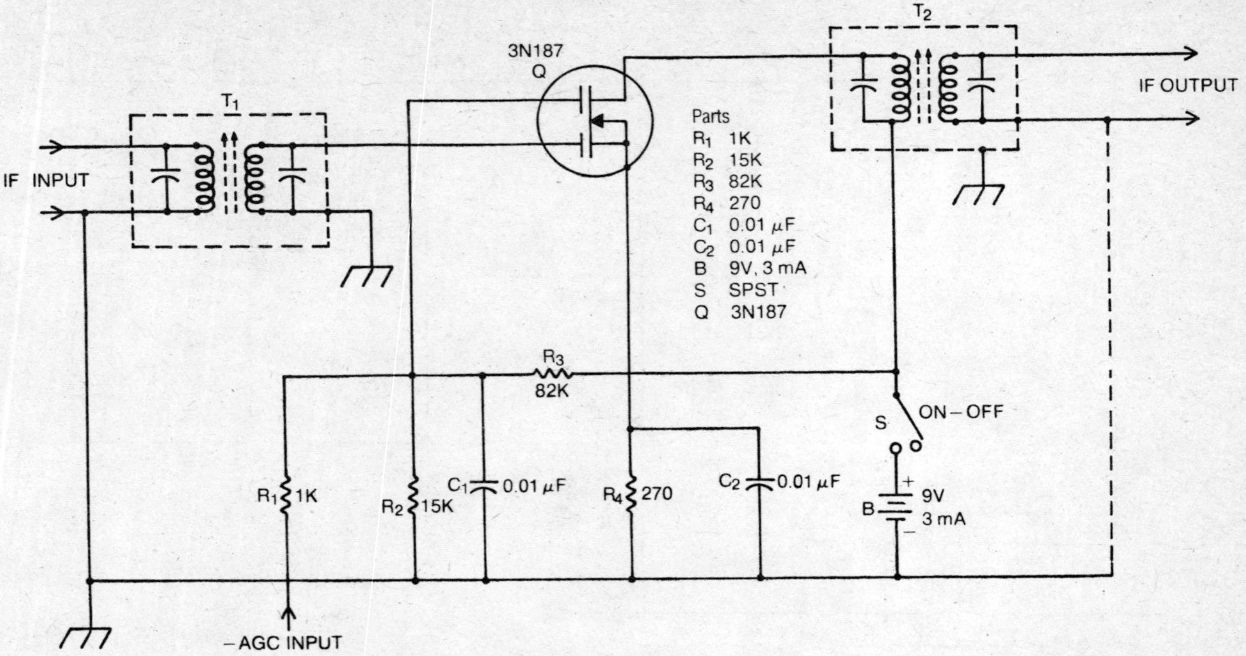

Fig. 3-10. Double-tuned IF amplifier.

either low or high frequencies (e.g., 455 kHz or 10.7 MHz). An advantage is the use of conventional, tube-type IF transformers (T_1 and T_2); a further advantage is that no neutralization is required, owing to the extremely small interelectrode capacitance of the MOSFET.

This single-stage amplifier is entirely conventional and requires no special explanation of theory or of adjustment. While all leads must be run as straight and short as practicable, no special gimmicks are necessary. Shielding is not required except that provided in the IF transformers themselves.

In this circuit, MOSFET Q receives its negative gate-1 bias from the voltage drop resulting from the flow of drain current through source resistor R_4, and its positive gate-2 bias from voltage divider R_2–R_3. With an individual MOSFET, resistance R_3 may require some adjustment for maximum IF gain. Current drain from the 9V source is approximately 3 mA, although this may vary somewhat with individual MOSFETs.

For increased gain and selectivity, two or three stages identical with the one shown in Fig. 3-10 may be cascaded with little difficulty, provided the usual decoupling *RC* filters are employed.

REGENERATIVE IF AMPLIFIER

Figure 3-11 shows a conventional double-tuned IF amplifier to which regeneration has been added for increased sensitivity and selectivity. Like the preceding circuit, this one also may be used at either low or high frequencies (e.g., 455 kHz or 10.7 MHz), the IF transformers (T_1 and T_2) being chosen accordingly. No neutralization is required, owing to the extremely small interelectrode capacitance of the MOSFET.

To obtain a simple regenerative circuit, use is made of an input IF transformer T_1 having a centertapped secondary (such a commercial unit for 455 kHz is the Miller 1312-C3). The matching output IF transformer T_2 has untapped windings (Miller 1312-C2). The tapped coil provides a Hartley-type oscillator circuit with the gate-1 circuit of the MOSFET. Regeneration is controlled by varying the positive gate-2 bias with regeneration control potentiometer R_3. When this potentiometer is adjusted for operation just below the point of oscillation, the sensitivity of the circuit is increased markedly, especially for voice signals; when the adjustment is taken slightly beyond the threshold point, the circuit will oscillate on CW signals, making their reception possible without a separate beat-frequency oscillator.

The regenerative IF amplifier is particularly useful in receivers in which only a single IF stage is desired and in

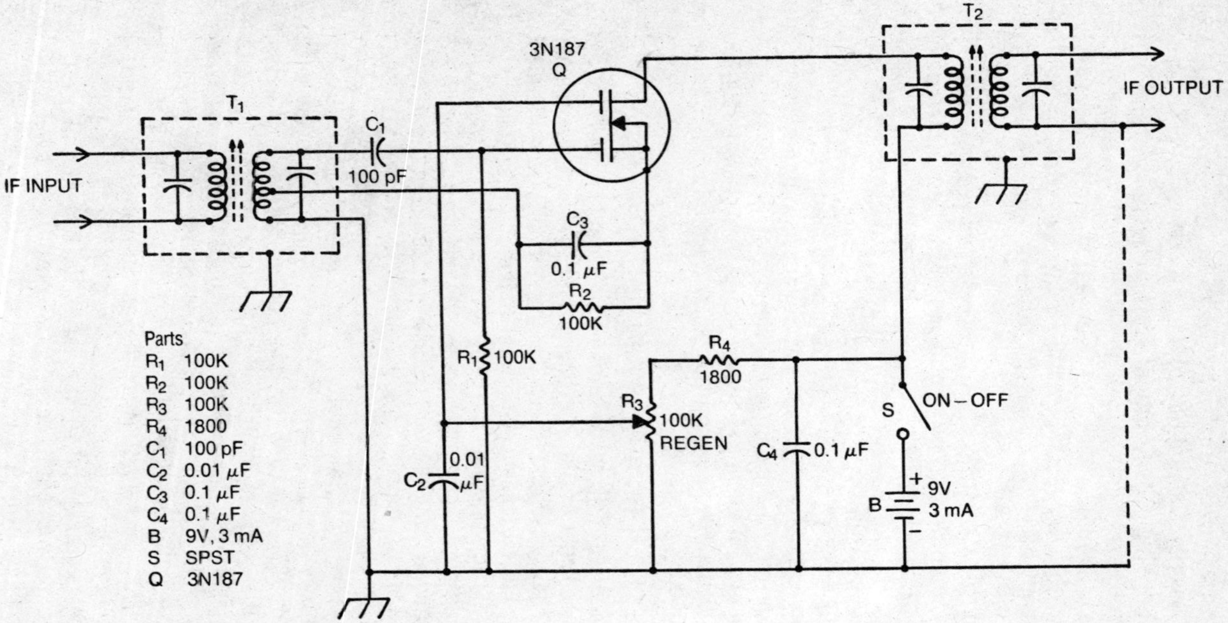

Parts

R_1	100K
R_2	100K
R_3	100K
R_4	1800
C_1	100 pF
C_2	0.01 μF
C_3	0.1 μF
C_4	0.1 μF
B	9V, 3 mA
S	SPST
Q	3N187

Fig. 3-11. Regenerative IF amplifier.

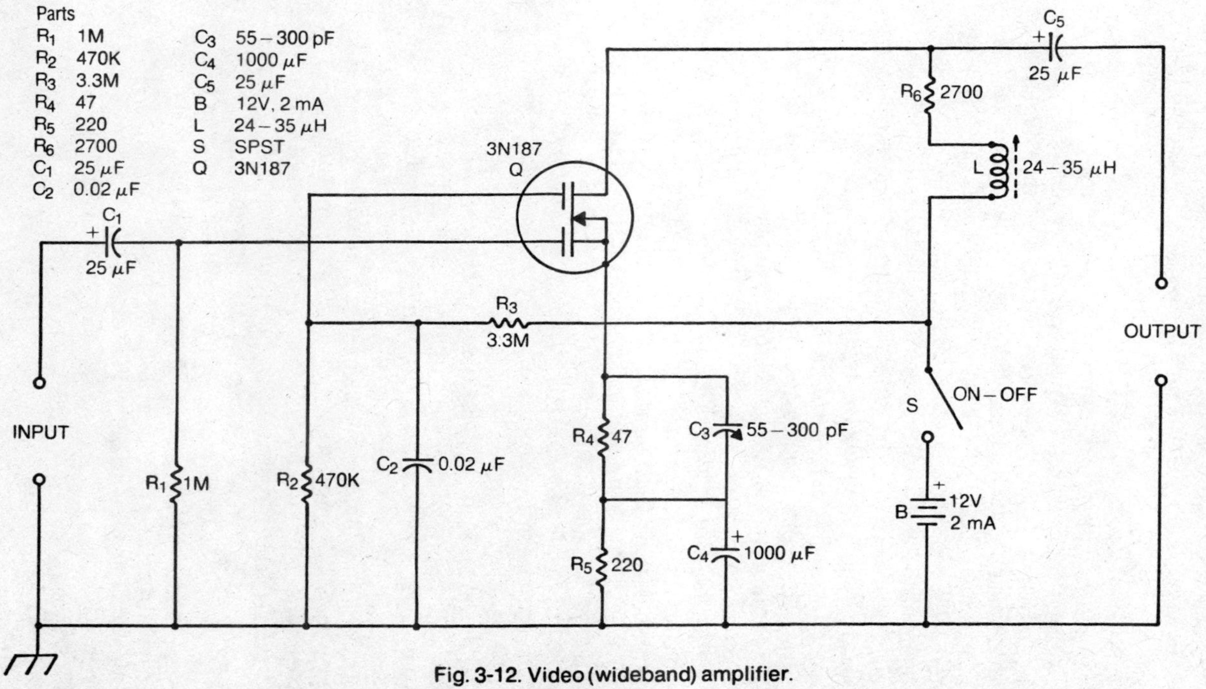

Fig. 3-12. Video (wideband) amplifier.

which general-purpose requirements dictate that both AM and CW signals be received.

Current drain from the 9V source is approximately 3 mA, although this may vary somewhat with individual MOSFETs.

VIDEO (WIDEBAND) AMPLIFIER

Video amplifiers are simple-appearing, principally *RC*-coupled circuits which exhibit wideband response, operating from low audio frequencies well into the high radio frequencies. Such amplifiers find a multitude of uses in electronics, but are known principally for their role in TV picture channels, oscilloscope horizontal and vertical channels, and instrument amplifiers (especially AC electronic-voltmeter boosters). Figure 3-12 shows the circuit of a simple MOSFET video amplifier.

Frequency response of the circuit extends from 60 Hz to 10 MHz. The input impedance is approximately 1M, determined principally by the resistance of the input resistor R_1. Output impedance is approximately 2800Ω. These input and output values hold at 1 kHz. The open-circuit voltage gain of the circuit is approximately 10 at 1 kHz, and is down approximately 3 dB at 50 Hz and down approximately 6 dB at 10 MHz. The maximum input signal voltage before output peak clipping is approximately 0.1V RMS, and the corresponding maximum output signal voltage is approximately 1V RMS.

The peaking elements are trimmer capacitor C_3 (55–300 pF) and slug-tuned coil L (24–35 μH, Miller 4508 or equivalent). Both of these components are adjusted carefully for maximum gain of the amplifier at 10 MHz. In a circuit of this type, successful high-frequency operation demands that all wiring be short, rigid, and direct, and that leads be dressed for optimum high-frequency performance.

In this circuit, MOSFET Q receives its negative gate-1 bias from the voltage drop resulting from the flow of drain current through source resistors R_4 and R_5 in series, and its positive gate-2 bias from voltage divider R_2–R_3, RF-bypassed by capacitor C_2. With an individual MOSFET, resistance R_3 may need some adjustment for maximum gain at 1 kHz. Current drain from the 12V source is approximately 2 mA, but this may vary somewhat with individual MOSFETs.

Chapter 4

Control Circuits and Devices

The high input resistance, high transconductance, and low temperature drift of the MOSFET recommend this type of transistor for use in control circuits and adjunct devices: sensitive electronic relays, interval timers, alarm devices, automation accessories, and so on. This chapter presents 20 such circuits, and these will suggest others to the alert experimenter.

In each of the circuits, unless indicated otherwise on the circuit diagram or in the text, capacitances are in picofarads, resistances in ohms, and inductances in henrys. Resistors are ½W, and electrolytic capacitors are 25V. For simplicity, batteries are shown for DC supply; however, a *well filtered* power supply can be used instead of a battery.

Before undertaking the wiring and operation of any circuit, read carefully the hints and precautions given in Chapter 1.

ZERO-CURRENT DC RELAY

Figure 4-1 shows the circuit of a sensitive DC relay which draws so little current (0.15 μA at 1.5V) that, for all practical purposes, it may be considered a zero-current device. The DC signal power required to close the relay is only 0.225 μW.

The circuit consists of a single-MOSFET DC amplifier with a milliampere-type DC relay connected in a 4-arm bridge circuit in the drain output section. The bridge allows the static drain current of the MOSFET to be balanced out of the relay (by adjustment of potentiometer R_3), and its four arms consist of resistor R_2, the two "halves" of potentiometer R_3, and the internal drain-to-source resistance of the MOSFET. The relay is a 1 mA, 1000Ω DC device (Sigma 5F-1000 or equivalent).

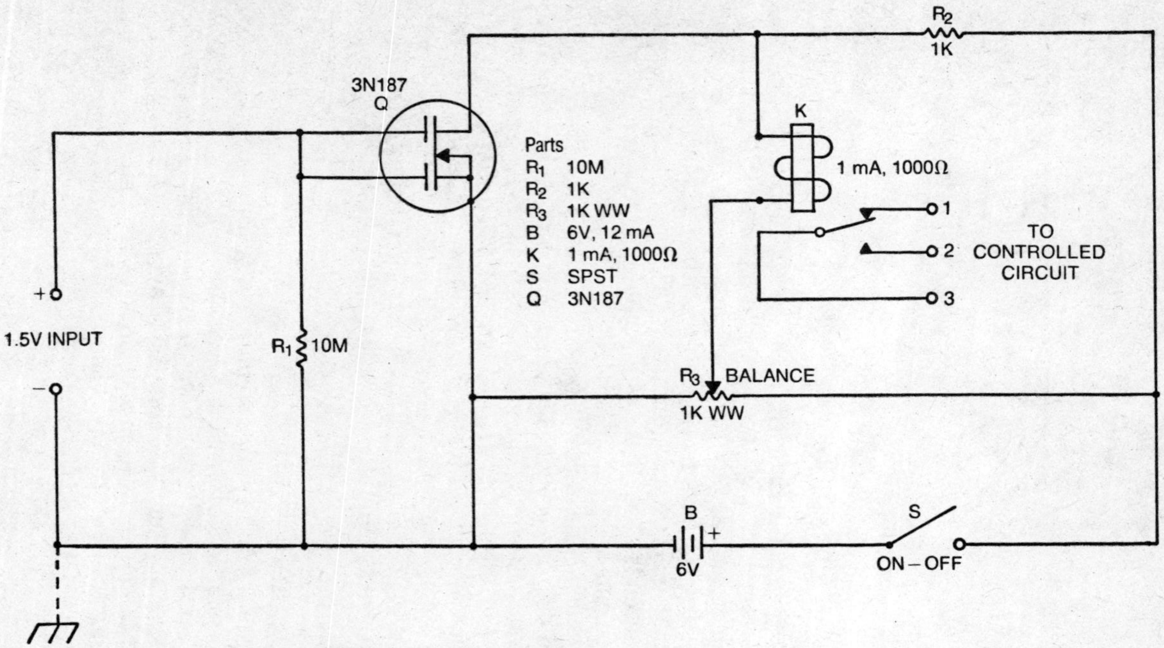

Fig. 4-1. "Zero-current" DC relay.

With zero signal input, balance control potentiometer R_3 is set to the point at which the relay opens. The circuit then will remain balanced unless the setting of R_3 is later disturbed. When a 1.5V DC signal subsequently is applied to the input terminals, the relay will close as a result of the consequent unbalance of the bridge circuit. The input resistance of the circuit (10M) is determined by gate-to-ground resistor R_1. For smaller input signal current than 0.15 μA, increase the resistance of R_1 proportionately. Use output terminals *1* and *3* for operations in which the external controlled circuit must be closed by the relay closure; use 2 and *3* for operations in which the external controlled circuit must be opened.

When the relay is closed, current drain from the 6V supply is approximately 12 mA, but this may vary somewhat with individual MOSFETs.

AC RF RELAY

The sensitive relay circuit shown in Fig. 4-2 operates on AC control signals of reasonably low amplitude. This arrangement consists of a relay-actuating DC amplifier preceded by a shunt diode rectifier. For relay closure, an AC input voltage of 1.1V RMS is required. The input resistance is nearly 0.5M. (Owing to the high reverse resistance—10M at −10V—of the 1N811 silicon diode *D*, high-value gate-to-ground resistor R_1 can be used.)

The shunt diode rectifier circuit (C_1–D–R_1) applies to the gates of the MOSFET a positive DC voltage equal approximately to the peak value of the AC input signal voltage; this is approximately +1.5V when the AC input signal is 1.1V RMS, and this is sufficient to close the relay. The junction capacitance of the 1N811 silicon diode limits efficient performance of the rectifier to the range extending from power-line frequencies to the supersonic frequencies. For RF operation into the tens of megahertz, use a 1N34A point-contact germanium diode; but this will necessitate reducing resistance R_1 to 10K or less.

The MOSFET portion of the circuit provides a DC amplifier with a milliampere-type DC relay connected in a 4-arm bridge circuit in the drain output section. The bridge allows the static drain current of the MOSFET to be balanced out of the relay (by adjustment of potentiometer R_3); its four arms consist of resistor R_2, the two "halves" of potentiometer R_3, and the internal drain-to-source resistance of the MOSFET. The relay is a 1 mA, 1000Ω DC device (Sigma 5F-1000 or equivalent).

With zero-signal input, balance control potentiometer R_3 set to the point at which the relay opens. The circuit then will remain balanced unless the setting of R_3 is later disturbed.

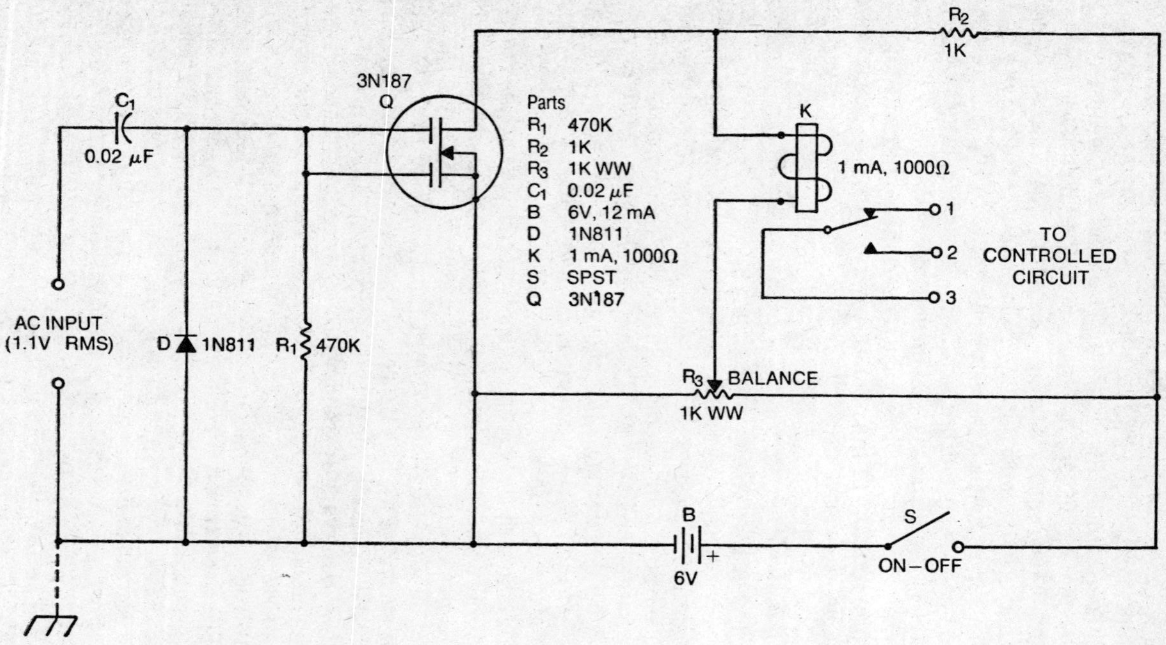

Fig. 4-2. AC–RF relay.

When a 1.1V AC or RF signal subsequently is applied to the input terminals, the relay closes as a result of the consequent unbalancing of the bridge circuit. Use output terminals *1* and *3* for operations in which the external controlled circuit must be closed by the relay closure; use *2* and *3* for operations in which the external controlled circuit must be opened.

When the relay is closed, current drain from the 6V supply is approximately 12 mA, but this may vary somewhat with individual MOSFETs.

SENSITIVE AF RELAY

Figure 4-3 shows the circuit of an audio-frequency AC relay which operates on an input-signal voltage of 0.1V RMS. This arrangement consists of a high-impedance-input AC voltage amplifier Q_1, shunt-diode rectifier $C_3–D–R_6$, and relay-actuating DC amplifier Q_2. The input AC amplifier has a voltage gain of approximately 10, so that this stage presents a signal voltage of approximately 1V RMS to the diode circuit. The diode circuit, in turn, presents to the DC amplifier a positive voltage equal approximately to the peak value of the 1V RMS output of Q_1, or approximately +1.4V. This is sufficient to close the relay. The input impedance of the circuit (determined principally by the full resistance of sensitivity control potentiometer R_1) is 0.5M.

In the AC amplifier stage, MOSFET Q_1 receives its negative gate-1 bias from the voltage drop resulting from the flow of drain current through source resistor R_5, and its positive gate-2 bias from voltage divider $R_2–R_3$. With an individual MOSFET Q_1, resistance R_3 may require some adjustment for maximum voltage gain in the input stage. In the DC amplifier stage, MOSFET Q_2 drives a milliampere-type DC relay connected in a 4-arm bridge circuit in the drain output section. The bridge allows the static drain current of Q_2 to be balanced out of the relay (by adjustment of potentiometer R_8); its four arms consist of resistor R_7, the two "halves" of potentiometer R_8, and the internal drain-to-source resistance of MOSFET Q_2. The relay is a 1 mA, 1000Ω DC device (Sigma 5F-1000 or equivalent).

With zero signal input at the input terminals, balance control potentiometer R_8 is set to the point at which the relay opens. The circuit then will remain balanced unless the setting of R_8 is later disturbed. With potentiometer R_1 set for maximum sensitivity, when a 0.1V RMS signal subsequently is applied to the input terminals, the relay closes as the result of the consequent unbalancing of the bridge circuit. Use output terminals *1* and *3* for operations in which the external

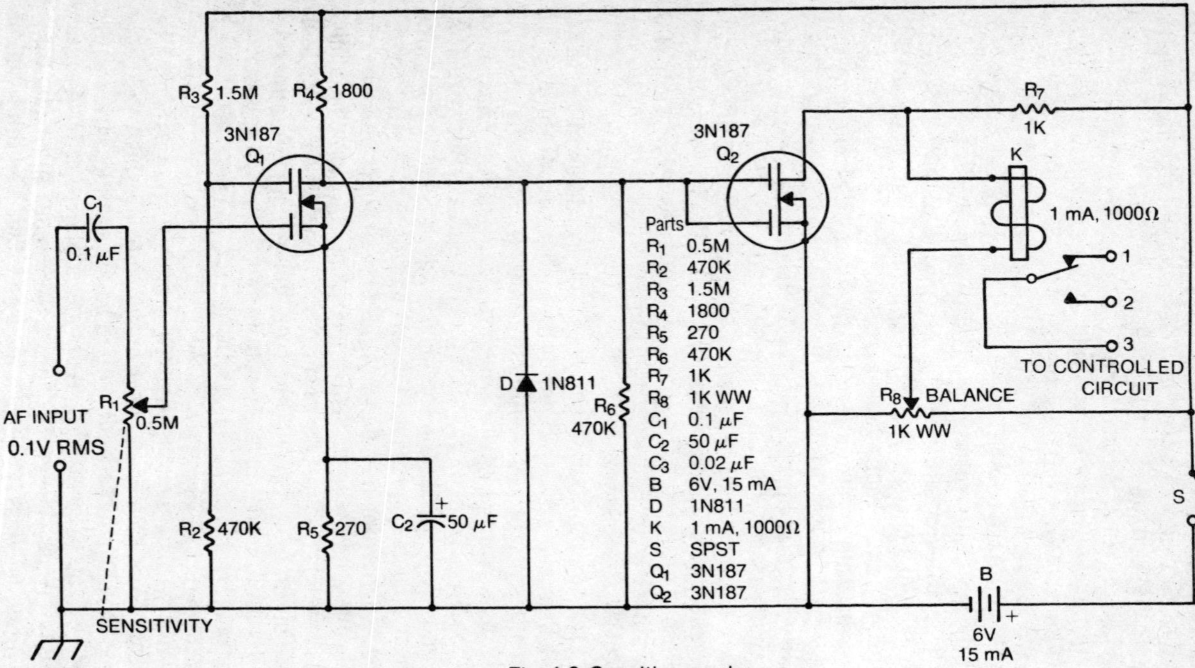

Parts	
R_1	0.5M
R_2	470K
R_3	1.5M
R_4	1800
R_5	270
R_6	470K
R_7	1K
R_8	1K WW
C_1	0.1 µF
C_2	50 µF
C_3	0.02 µF
B	6V, 15 mA
D	1N811
K	1 mA, 1000Ω
S	SPST
Q_1	3N187
Q_2	3N187

Fig. 4-3. Sensitive AF relay.

controlled circuit must be closed by the relay closure; use 2 and 3 for operations in which the external controlled circuit must be opened.

When the relay is closed, current drain from the supply is approximately 15 mA, but this may vary somewhat with individual MOSFETs.

LC-TUNED AF RELAY

Relay operation often is desired at a single frequency. In Fig. 4-4, single-frequency operation is achieved by tuning the AC amplifier stage of a circuit which is substantially the same in other particulars as the one described in the preceding section. Tuning is accomplished with parallel-resonant circuit $L-C_2$ in the drain circuit of MOSFET Q_1. The AC output of the tuned amplifier is rectified by the shunt diode circuit C_5-D-R_5, and the resulting DC voltage is applied to DC amplifier Q_2, which, in turn, actuates milliampere-type DC relay K.

Inductor L and capacitor C_2 must be chosen for the desired frequency f_r. Generally, the inductor will govern the caacitance, since few inductors for AF use can be varied. With a given inductance L, the required capacitance C_2 which can be built up accurately can be determined with the aid of the following equation, in which L is in henrys, C_2 in farads, and f_r in hertz:

$$C_2 = \frac{1}{39.5 f_r^2 L} \quad \text{(Eq. 4-1)}$$

From this equation, it is seen that for an available 10H inductor, the required capacitance C_2 for resonance at 1000 Hz is 0.0025 μF.

The input AC amplifier Q_1 has a voltage gain of approximately 10, so that this stage presents an amplified signal voltage of 1V RMS to the diode circuit when the input signal voltage at the input terminals is 0.1V RMS. The diode circuit, in turn, presents to the DC amplifier Q_2 a positive voltage equal approximately to the peak value of the 1V RMS output of Q_1, or approximately +1.4V. This is sufficient to close relay K.

In the DC amplifier stage, MOSFET Q_2 drives a milliampere-type DC relay in a 4-arm bridge circuit in the drain output section. The bridge allows the static drain current of Q_2 to be balanced out of the relay (by adjustment of potentiometer R_6), and its four arms are resistor R_7, the two "halves" of potentiometer R_6, and the internal drain-to-source resistance of MOSFET Q_2. The relay is a 1 mA, 1000Ω DC device (Sigma 5F-1000 or equivalent).

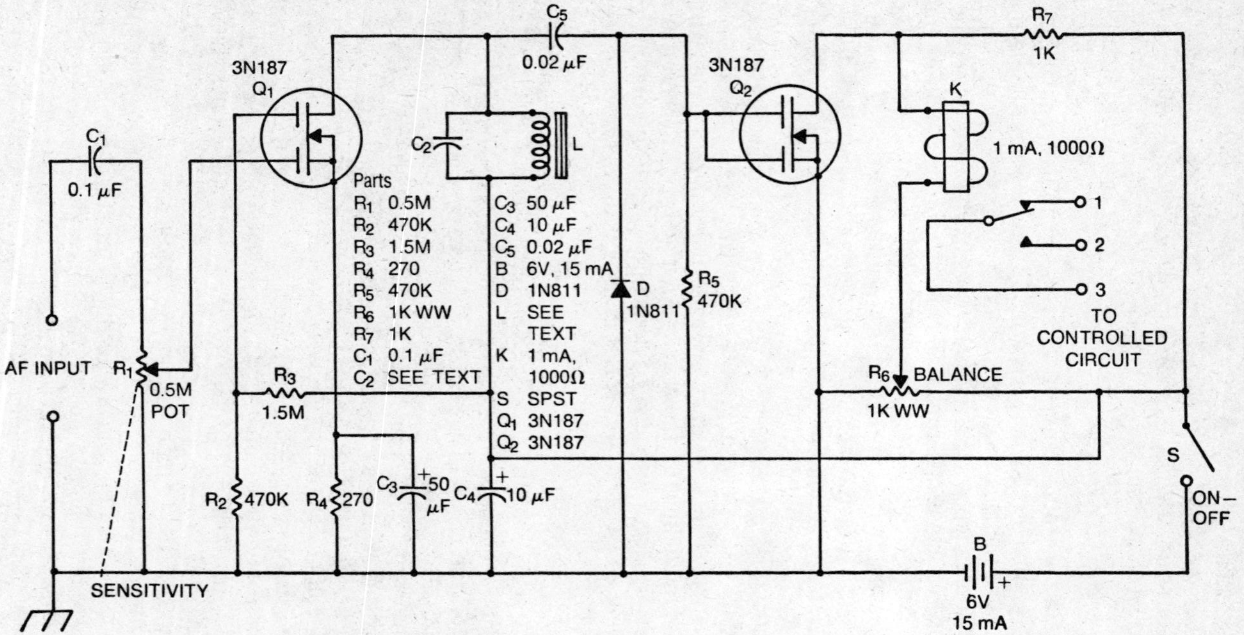

Fig. 4-4. LC-tuned AF relay.

With zero-signal input, balance control potentiometer R_6 is set to the point at which the relay opens. The circuit then will remain balanced unless the setting of R_6 is later disturbed. When a 0.1V AC signal at the frequency determined by L and C_2 then is applied to the input terminals (with potentiometer R_1 set for maximum sensitivity), the relay closes as a result of the consequent unbalancing of the bridge circuit. Use output terminals *1* and *3* for operations in which the external controlled circuit must be closed by the relay closure; use 2 and *3* for operations in which the external controlled circuit must be opened.

When the relay is closed, current drain from the 6V supply is approximately 15 mA, but this may vary somewhat with individual MOSFETs.

RC-TUNED AF RELAY

The tuned audio-frequency relay circuit shown in Fig. 4-5 performs in the same manner as the *LC*-tuned circuit described in the preceding section. The *RC*-tuned circuit, however, needs no inductor, but achieves audio-frequency tuning by means of a parallel-*T* network (C_2–C_3–C_4–R_6–R_7–R_8) in the input stage of the circuit. In all other respects, the circuit is similar to the one shown earlier in Fig. 4-4. Resistance–capacitance tuning is of value when iron-core inductors are to be avoided and when continuously variable tuning is desired.

In this circuit, tuning to a peak is accomplished by incorporating a *null* circuit (the parallel-*T* network) in the negative feedback loop between the drain output and gate-1 input section of the first stage (Q_1). The result of this is that negative feedback occurs at all frequencies except the null frequency of the network (null frequency $f_r = 1/6.28R_6C_2$ in Fig. 4-5), which is removed by the network, canceling the gain at those frequencies. But the amplifier has full gain at frequency f_r, since this one frequency is not included in the feedback. The net result is tuned bandpass-peak response.

In the *RC* network, $C_2 = C_3 = 0.5C_4$, and $R_6 = R_7 = 2R_8$. When these relationships are preserved, the null frequency is calculated as

$$f_r = \frac{1}{6.28R_6C_2} \qquad \text{(Eq. 4-2)}$$

where f_r is in hertz, R_6 in ohms, and C_2 in farads.

From Eq. 4-2, it is seen that for a parallel-*T* network in which C_2 and C_3 are each 0.1 μF, $C_4 = 2C_3 = 0.2\ \mu$F, R_6 and R_7 are each 1592Ω, and $R_8 = 0.5R_6 = 796\Omega$, the null frequency is

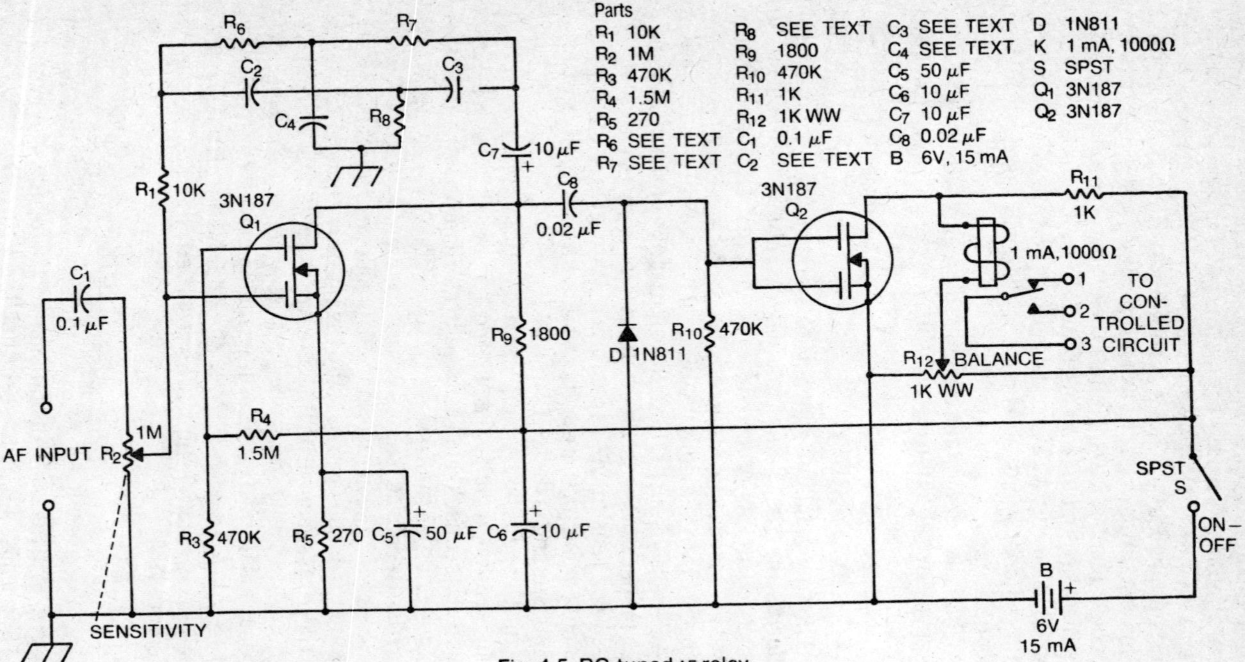

Fig. 4-5. RC-tuned AF relay.

1000 Hz (the peak frequency f_r of the amplifier). In general, it is best to select the three capacitances ($C_2 = C_3 = 0.5C_4$) and then to adjust the resistances to exact values ($R_6 = R_7 = 2R_8$) as required. The following equations—in which resistances are in ohms, capacitances are in farads, and peak frequency is in hertz—will help:

$$R_6 = R_7 = \frac{1}{6.28 f_r C_2} \quad \text{(Eq. 4-3)}$$

$$R_8 = 0.5R_6 \quad \text{(Eq. 4-4)}$$

The resistances and capacitances must have exact values. The sharpness of tuning will depend upon the closeness with which these components meet specified values. The circuit may be made continuously tunable by using a 3-gang potentiometer for R_6–R_7–R_8 and switching the capacitors in groups of three to change frequency ranges. In this way, the amplifier can be given tuning ranges of 20–200 Hz, 200–2000 Hz, and 2–20 kHz, to cover the audio-frequency spectrum.

Input AC amplifier Q_1 has a voltage gain of approximately 10, so that this stage presents an amplified signal voltage of 1V RMS to the diode circuit C_8–D–R_{10} when the signal voltage at the input terminals is 0.1V RMS and potentiometer R_2 is set for maximum sensitivity. The diode circuit, in turn, rectifies this voltage and presents a resulting DC voltage to DC amplifier Q_2. This latter voltage is approximately +1.4V, which is the peak value of the 1V RMS output of the first stage. This is sufficient to close the relay.

In the DC amplifier stage, MOSFET Q_2 drives a milliampere-type DC relay in a 4-arm bridge circuit in the drain output section. The bridge allows the static drain current of Q_2 to be balanced out of the relay (by adjustment of potentiometer R_{12}), and its four arms consist of resistor R_{11}, the two "halves" of potentiometer R_{12}, and the internal drain-to-source resistance of MOSFET Q_2. The relay is a 1 mA, 1000Ω device (Sigma 5F-1000 or equivalent).

When the relay is closed, current drain from the 6V supply is approximately 15 mA, but this may vary somewhat with individual MOSFETs.

COINCIDENCE RELAY

A relay that closes when two signals arrive at the same instant (or when they overlap for a part of their duration) is useful in various counting, sampling, and automatic control systems. Figure 4-6 shows a *coincidence* relay of this type, employing a single MOSFET. A MOSFET in this application presents high resistance (low loading) to both signal sources.

Parts

R_1	1M
R_2	1M
R_3	1K
R_4	1K WW
B	12V, 24 mA
K	2 mA, 2500Ω
S	SPST
Q	3N187

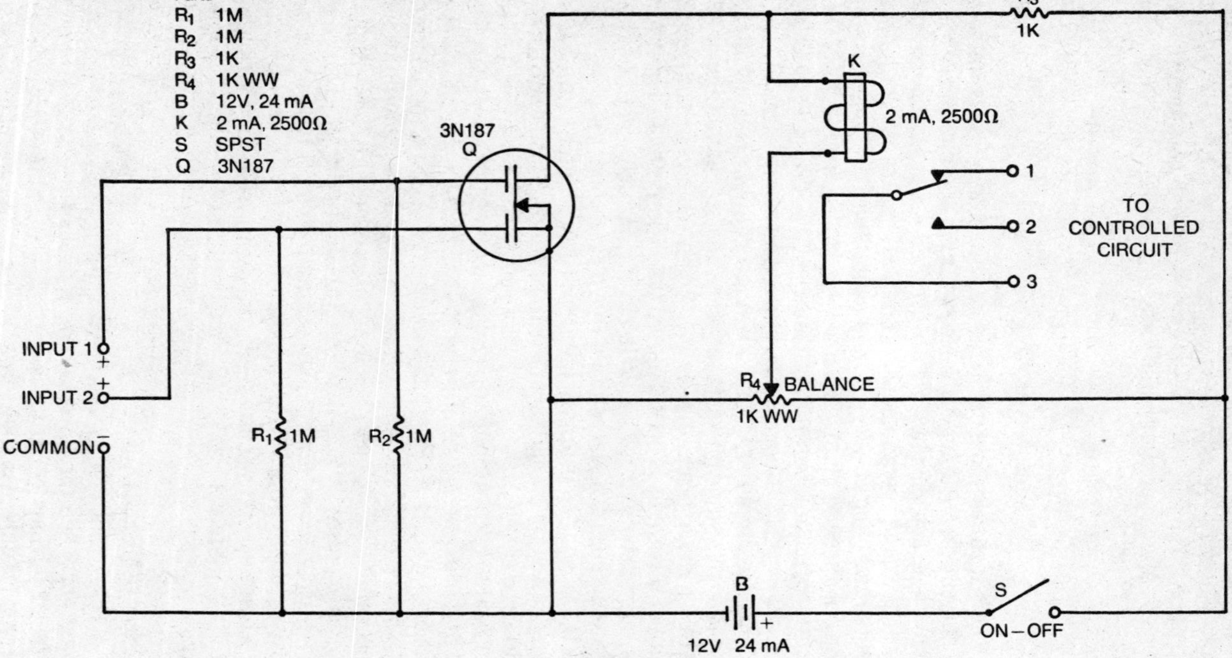

Fig. 4-6. Coincidence relay.

Each input signal has a voltage of +1.5V. Signal 1 is applied to gate 2 of the MOSFET, and signal 2 is applied to gate 1. When either one of these signals is applied alone, the drain-current increase that it produces is insufficient to close the relay. When both signals are applied at the same time, however, the increase in drain current will do so.

The circuit is essentially that of a DC amplifier. The relay operates in a 4-arm bridge in the drain circuit. The bridge allows the static drain current to be balanced out of the relay (by adjustment of potentiometer R_4); its four arms consist of resistor R_3, the two "halves" of potentiometer R_4, and the internal drain-to-source resistance of MOSFET Q. The relay is a 2 mA, 2500Ω device (Sigma 5F-2500 or equivalent). With zero signal at input 1 and input 2, balance control potentiometer R_4 is set to the point at which the relay opens. The circuit then will remain balanced unless the setting of R_4 is later disturbed. Use output terminals *1* and *3* for operations in which the external controlled circuit must be closed by the relay closure; use 2 and *3* for operations in which the external controlled circuit must be opened.

When the relay is closed, current drain from the 12V supply is approximately 24 mA, but this may vary somewhat with individual MOSFETs.

SENSITIVE PHOTOELECTRIC RELAY

The output of a conventional selenium self-generating photocell (typically 0.3V DC, 77 μA, at 100 footcandles) is too low to operate a DC relay directly. The circuit given in Fig. 4-7 provides a simple DC amplifier for boosting the output of such a photocell. In this circuit, the DC output of photocell *PC* is applied directly to the MOSFET (positive output to the gates connected in parallel, and negative output to the source electrode), and a milliampere-type DC relay is operated in a 4-arm bridge in the drain circuit.

The bridge—consisting of resistor R_2, the two "halves" of potentiometer R_1, and the internal drain-to-source resistance of the MOSFET—allows the static drain current to be balanced out of the relay (by adjustment of potentiometer R_1). The relay, K, is a 1 mA, 1000Ω device (Sigma 5F-1000 or equivalent).

With the photocell darkened, balance control potentiometer R_1 is set to the point at which the relay opens. The circuit then will remain balanced unless the setting of R_1 is later disturbed. When the photocell subsequently is illuminated, the relay will close. Use output terminals *1* and *3* for operations in which the external controlled circuit must be

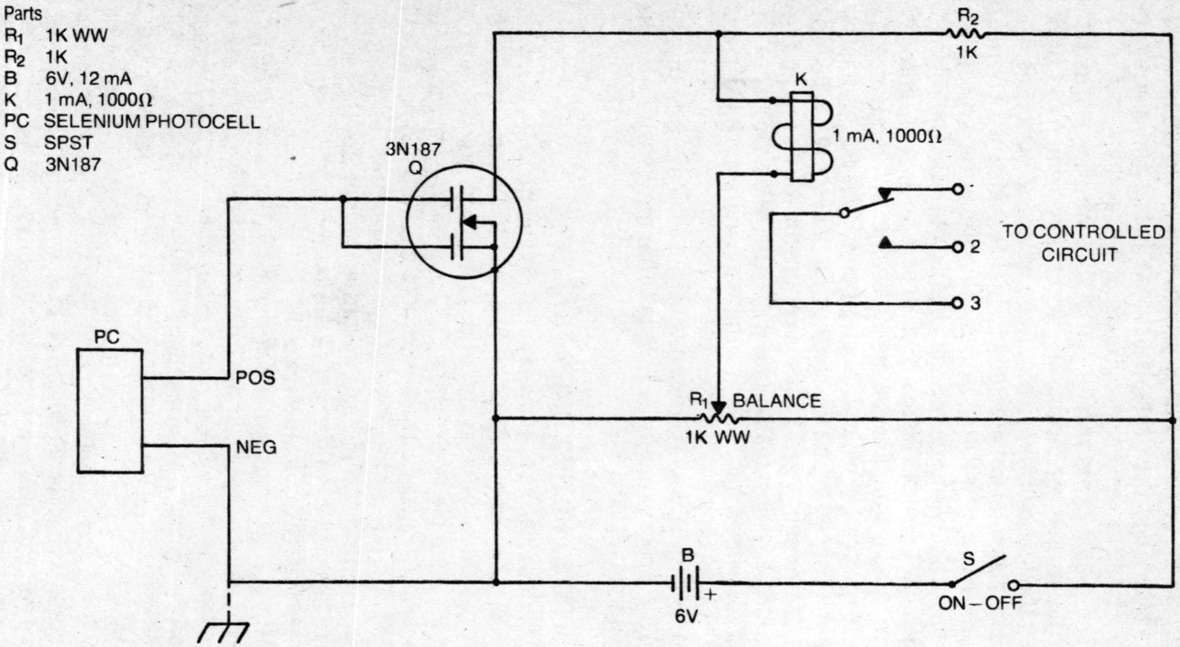

Fig. 4-7. Sensitive photoelectric relay.

closed by the relay closure; use 2 and *3* for operations in which the external controlled circuit must be opened.

When the relay is closed, current drain from the 6V supply is approximately 12 mA, but this may vary somewhat with individual MOSFETs.

TEMPERATURE-SENSITIVE RELAY

Figure 4-8 shows the circuit of an electronic temperature-sensitive relay. In this arrangement, a thermistor (thermally sensitive resistor) R_1 is operated from the 6V DC supply along with a simple DC amplifier based on the MOSFET Q. The DC output of the thermistor is applied to the MOSFET to close the relay.

The thermistor and potentiometer R_2 form a voltage divider which allows close selection of the temperature-dependent voltage applied to the gates. As the temperature rises, the resistance of the thermistor decreases proportionately, allowing the thermistor current to increase and the output of potentiometer R_2 (and accordingly the gate voltage) to increase. Potentiometer R_2 is adjusted to the point at which the gate voltage at the desired temperature is sufficient to close the relay.

Relay K operates in a 4-arm bridge—consisting of resistor R_4, the two "halves" of potentiometer R_3, and the internal drain-to-source resistance of the MOSFET—in the drain circuit. This bridge allows the static drain current to be balanced out of the relay (by adjustment of potentiometer R_3). The relay is a 1 mA, 1000Ω device (Sigma 5F-1000 or equivalent).

With the thermistor temporarily disconnected, balance control potentiometer R_3 is set to the point at which the relay opens. The circuit then will remain balanced unless the setting of R_3 later is disturbed. The thermistor may be a Fenwal GB42JM1 device, or the equivalent.

When the relay is closed, current drain from the 6V supply is approximately 14 mA.

TOUCH-PLATE RELAY

Figure 4-9 shows the circuit of a simple electronic relay that responds to a light touch of the finger. Such touch-plate relays have become common for operating all sorts of electrical equipment. In this arrangement, gate 1 of the MOSFET is grounded, and gate 2 is allowed to float and is connected to the pickup, which is a small disc of sheet metal or foil. The floating gate is highly susceptible to pickup, so that merely touching the finger to it couples in enough stray pickup energy to drive the MOSFET.

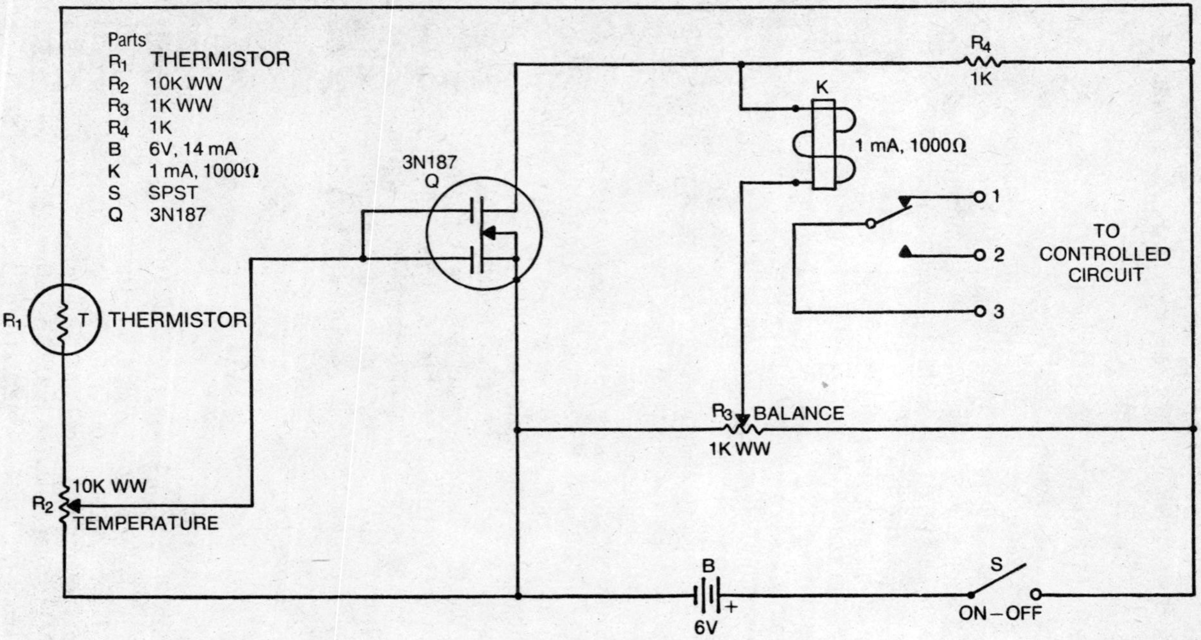

Fig. 4-8. Temperature-sensitive relay.

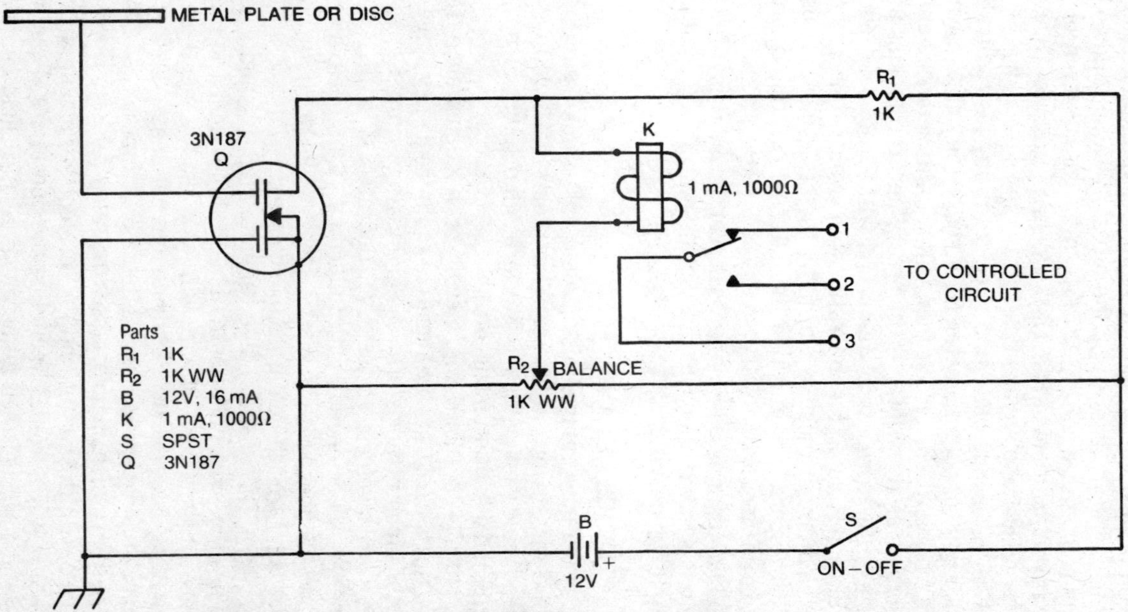

Fig. 4-9. Touch-plate relay.

The MOSFET serves as a simple DC amplifier with a milliampere-type DC relay in its drain circuit. The relay is connected in a 4-arm bridge circuit—consisting of resistor R_1, the two "halves" of potentiometer R_2, and the internal drain-to-source resistance of the MOSFET—which allows the static drain current of the MOSFET to be balanced out of the relay (by adjustment of R_2). The relay is a 1 mA, 1000Ω device (Sigma 5F-1000 or equivalent).

With the pickup plate protected from any nearby bodies, balance control potentiometer R_2 is set to the point at which the relay opens. The circuit then will remain balanced unless the setting of R_2 is later disturbed. When the pickup plate subsequently is touched, the relay will close. Use output terminals *1* and *3* for operations in which the external controlled circuit must be closed by the relay closure; use 2 and *3* for operations in which the external controlled circuit must be opened.

When the relay is closed, current drain from the 12V supply is approximately 16 mA, but this may vary somewhat with individual MOSFETs.

CAPACITANCE RELAY

Capacitance relays, also called *instrusion relays*, are familiar in many areas of electronics, as well as in nonelectronic applications. They are often found in burglar alarms and other anti-intrusion devices.

Figure 4-10 shows the circuit of a capacitance relay employing a MOSFET in the sensitive radio-frequency stage and a silicon bipolar transistor in the high-gain DC amplifier stage. In this arrangement, MOSFET Q_1 operates at a high radio frequency in conjunction with inductor *L*, trimmer capacitor C_2, and radio-frequency choke *RFC*. (All three of these components are self-contained in a commercial unit *T*, such as Miller 695.) The MOSFET stage oscillates lightly; and when a hand or other part of the human body comes near the pickup antenna (a short piece of wire with or without a metal disc or plate on its end), the MOSFET drain current undergoes a small change. This change is coupled to the base input of the 2N2712 DC amplifier which, in turn, closes the relay.

To set up the circuit initially, protect the pickup antenna from any nearby bodies, including the operator's. Then work back and forth between adjustments of threshold control R_5 and sensitivity control R_3 until relay *K* closes. Next, back off the setting of R_5 until the relay just opens. Then bring the hand close to the antenna and adjust trimmer C_2 and sensitivity control R_3 until the relay closes. Withdrawing the hand should

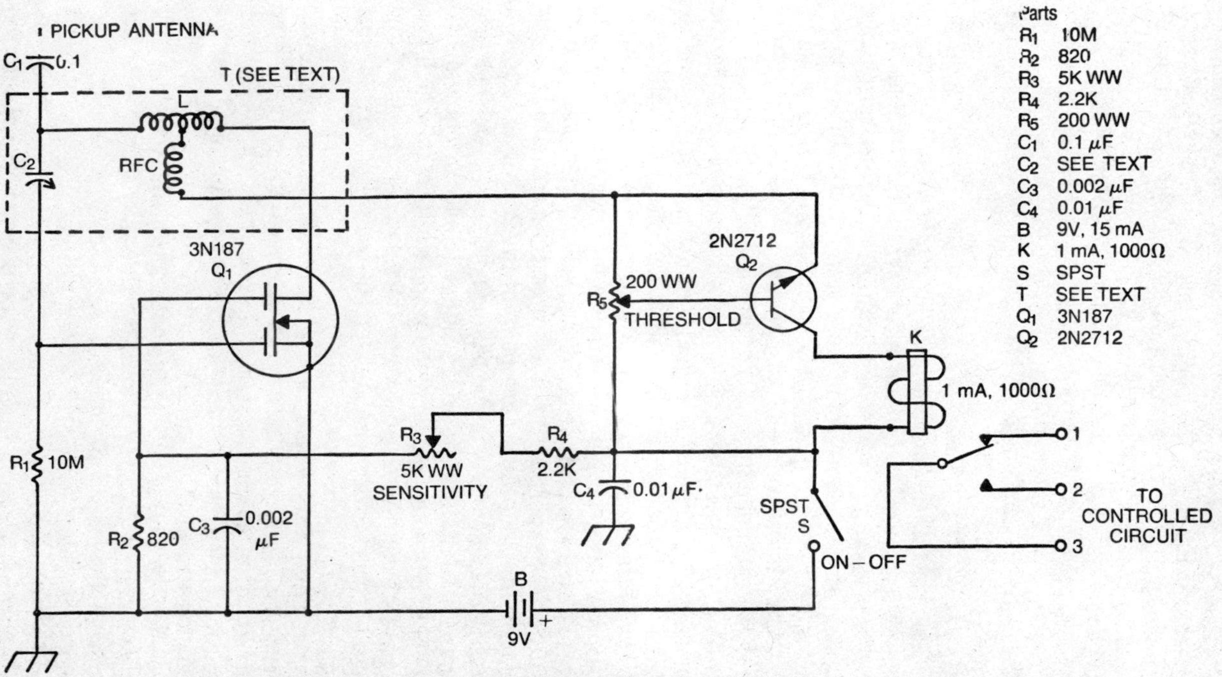

Fig. 4-10. Capacitance relay.

then cause the relay to open. When the circuit is correctly adjusted, bringing the hand to the desired distance from the antenna should close the relay, and withdrawing the hand should open it. The circuit must not be so adjusted that it will self-operate intermittently.

The relay is a 1 mA, 1000Ω device (Sigma 5F-1000 or equivalent). When the relay is closed, current drain from the 9V supply is approximately 15 mA, but this may vary somewhat with an individual 3N187 and 2N2712.

SOUND SWITCH

The circuit shown in Fig. 4-11 closes a relay when it "hears" a sound of the proper intensity. In this arrangement, the MOSFET acts as an AC voltage amplifier, diode D as a signal rectifier, and silicon bipolar transistor Q_2 as a DC amplifier driving relay K. A small crystal microphone *(MIC)* picks up the sound and delivers a proportionate AC voltage to gate 1 of the MOSFET through sensitivity control potentiometer R_1.

The MOSFET delivers an amplified sound voltage to the shunt rectifier circuit (C_3, D, and the internal base-to-emitter path of transistor Q_2), which, in turn, delivers the resulting DC voltage to the base of Q_2. The latter transistor then amplifies the DC and closes relay K. The diode must be poled as shown in Fig. 4-11, so that a positive DC voltage is applied to the Q_2 base.

The relay operates directly in the collector circuit of the 2N2712. Since the static current I_{co} of the 2N2712 is so low, it cannot affect the relay, so no balancing circuit is needed. The relay is a 1 mA, 1000Ω device (Sigma 5F-1000 or equivalent). Use output contacts *1* and *3* for operations in which the external circuit must be closed by the relay closure; use 2 and *3* for operations in which the external controlled circuit must be opened.

Initially, set sensitivity control R_1 for maximum sensitivity. Then, with the microphone picking up the desired level of noise, adjust R_5 so that the relay just closes. Louder intensities then may be accommodated by turning down the adjustment of R_1.

When the relay is closed, current drain from the 12V supply is approximately 17 mA, but this may vary somewhat with an individual 3N187 and 2N2712.

INTERVAL TIMER

Figure 4-12 shows the circuit of a conventional interval timer in which the very high input resistance of the MOSFET affords performance comparable to that of the usual vacuum-tube model. Because of the smallness of the MOSFET,

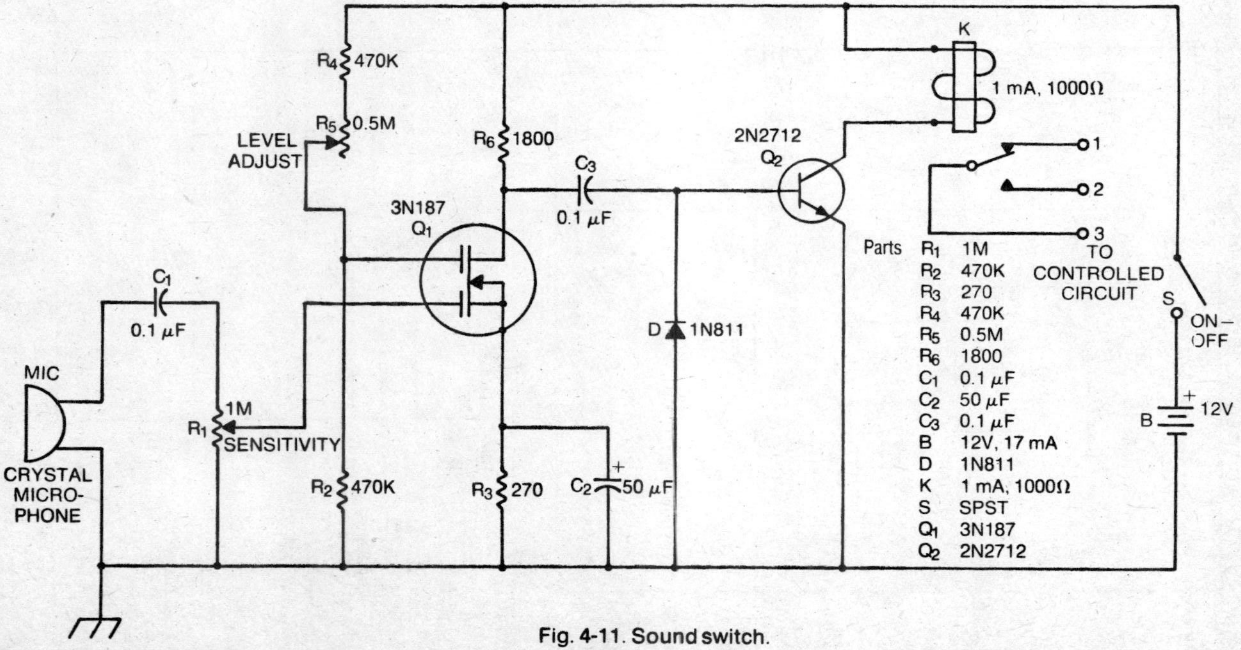

Fig. 4-11. Sound switch.

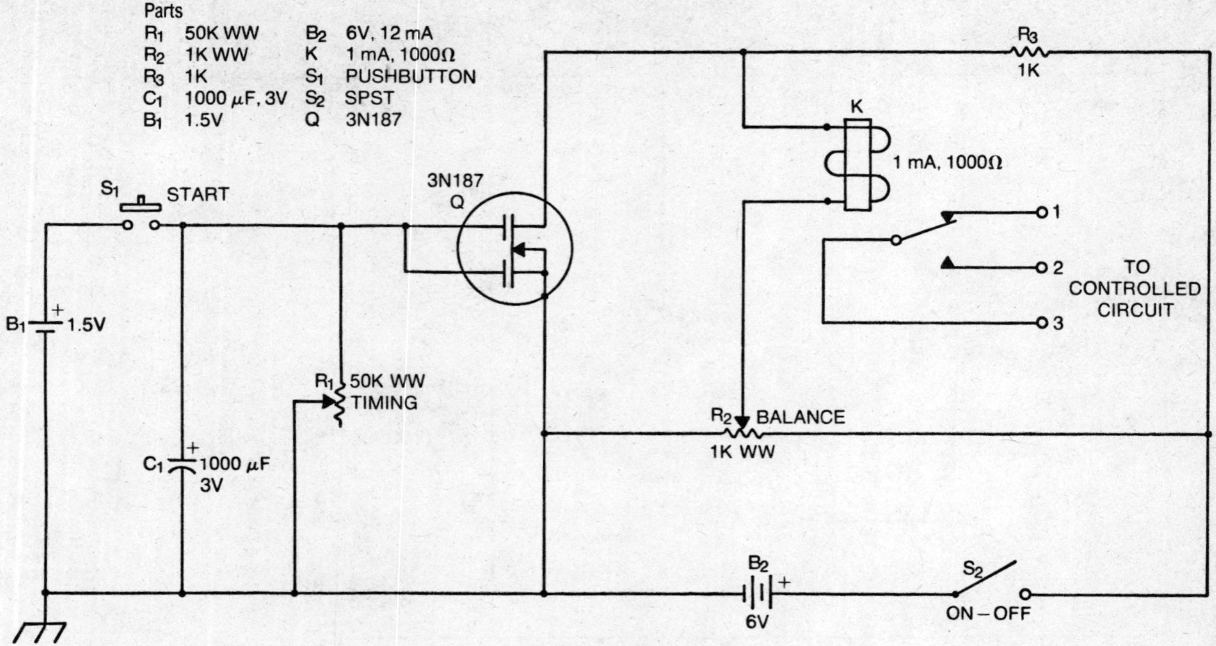

Fig. 4-12. Interval timer.

relay, 1000 μF (3V) capacitor, and other components, the entire unit can be very small and compact.

In this arrangement, the 3N187 MOSFET serves as a simple DC amplifier driving relay K in its drain circuit. When pushbutton switch S_1 is depressed momentarily, capacitor C_1 is charged from 1.5V cell B_1. At the same time, relay K is closed. When S_1 then is released, capacitor C_1 discharges at a rate determined by resistance R_1 and capacitance C_1 (the time interval before the relay drops out is closely equal to the time constant of the circuit: $t = RC$ seconds). The time instant that the relay drops out may be controlled by proper setting of timing control rheostat R_1. With the 50,000Ω control and 1000 μF capacitor shown, the maximum time interval is 50 sec. Other maximum values may be obtained by appropriate selection of C_1 and R_1 values in accordance with the time-constant formula.

The relay is a 1 mA, 1000Ω device (Sigma 5F-1000 or equivalent) operated in a 4-arm bridge in the drain circuit. The bridge—consisting of resistor R_3, the two "halves" of balance control potentiometer R_2, and the internal drain-to-source resistance of the MOSFET—allows the static drain current to be balanced out of the relay (by adjustment of potentiometer R_2) before attempting to operate the timer.

When the relay is closed, current drain from the 6V supply B_2 is approximately 12 mA, but this may vary somewhat with individual MOSFETs. The 1.5V cell, B_1, is used only intermittently, supplying charging current to capacitor C_1, and should enjoy long life, especially if a size D cell is used.

CONTINUOUSLY VARIABLE PHASE SHIFTER

Figure 4-13 shows the circuit of a device which will give smooth variation of phase between input and output of approximately zero to 180° by adjusting a single control (R_6). The circuit is basically that of a common-source amplifier with unbypassed source resistor (R_4).

When R_6 is adjusted to full resistance, it offers minimum loading of the MOSFET, and the output signal is effectively taken from the drain electrode and is 180° out of phase with the input signal. When instead, R_6 is adjusted to zero (potentiometer minimum), the output signal is effectively taken from the source electrode, somewhat in the manner of a source follower, and the output is of the same phase (0°) as the input. Between these two limits, a smooth variation of phase is obtained. Capacitor C_3 for output coupling must be chosen, with respect to the external load, such that it does not

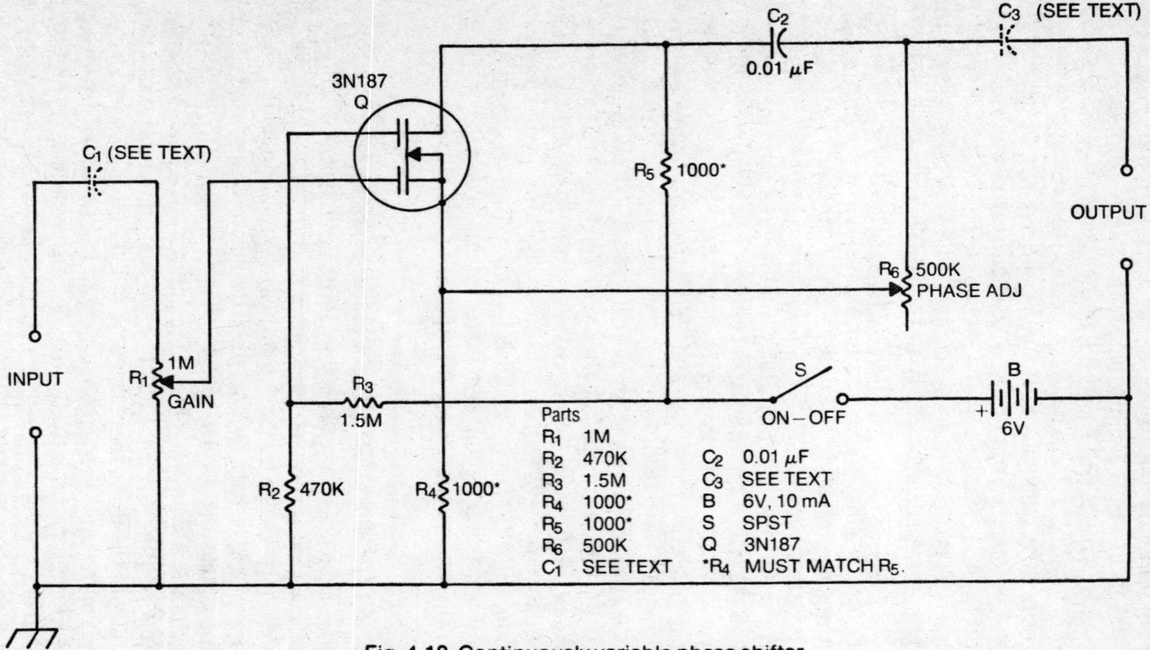

Fig. 4-13. Continuously variable phase shifter.

introduce additional undesired phase shift. This applies also to input capacitor C_1.

Voltage gain of the circuit is approximately 3.5. The maximum input signal voltage before peak clipping is 0.6V RMS; the corresponding output signal voltage is 2.1V RMS. In this circuit, gate 1 of the MOSFET receives its negative bias from the voltage drop resulting from the flow of drain current through source resistor R_4, and gate 2 receives its positive bias from voltage divider R_2–R_3. With an individual MOSFET, resistance R_3 may need some adjustment for maximum gain.

Current drain from the 6V source is approximately 10 mA, but this may vary somewhat with individual MOSFETs.

STEP-TYPE PHASE SHIFTER

The circuit shown in Fig. 4-14 is of the same type as the variable phase shifter described in the preceding section and shown in Fig. 4-13, except that here the phase is varied in steps by means of phase selector switch S_2, instead of being continuously variable. This difference will suit many applications calling for discrete steps of phase. A common capacitor (C_2, 0.01 μF) is employed for all phase steps. It is only necessary, then, to preset each of the potentiometers (R_6, R_7, R_8...R_n) for the desired phase shift, with the aid of an oscilloscope set up for Lissajous patterns. Each potentiometer can be 500,000Ω maximum.

The general particulars of operation of the circuit are the same as those given for the variable phase shifter in the previous section.

SENSITIVE CARRIER-FAILURE ALARM

A carrier-failure alarm is a useful monitor at radio stations of all kinds that remain on the air for extended periods. When amplification is employed in a device of this type, very little RF pickup is required, and no connection to the transmitter is necessary. Figure 4-15 shows a carrier-failure circuit embodying a diode detector and MOSFET DC amplifier. The latter operates a relay which actuates the alarm device itself.

A simple diode detector circuit is employed. This consists of tuned circuit L–C_1, 1N34A germanium diode D, load resistor R_1, bypass capacitor C_2, radio-frequency choke RFC, and bypass capacitor C_3. Capacitance C_1 and inductance L are selected to resonate at the frequency of the monitored carrier. Table 4-1 gives coil-winding instructions for coils to cover the frequency range 440 kHz to 30 MHz when C_1 is a 365 pF variable capacitor. For ham bands, use a 100 or 50 pF capacitor and

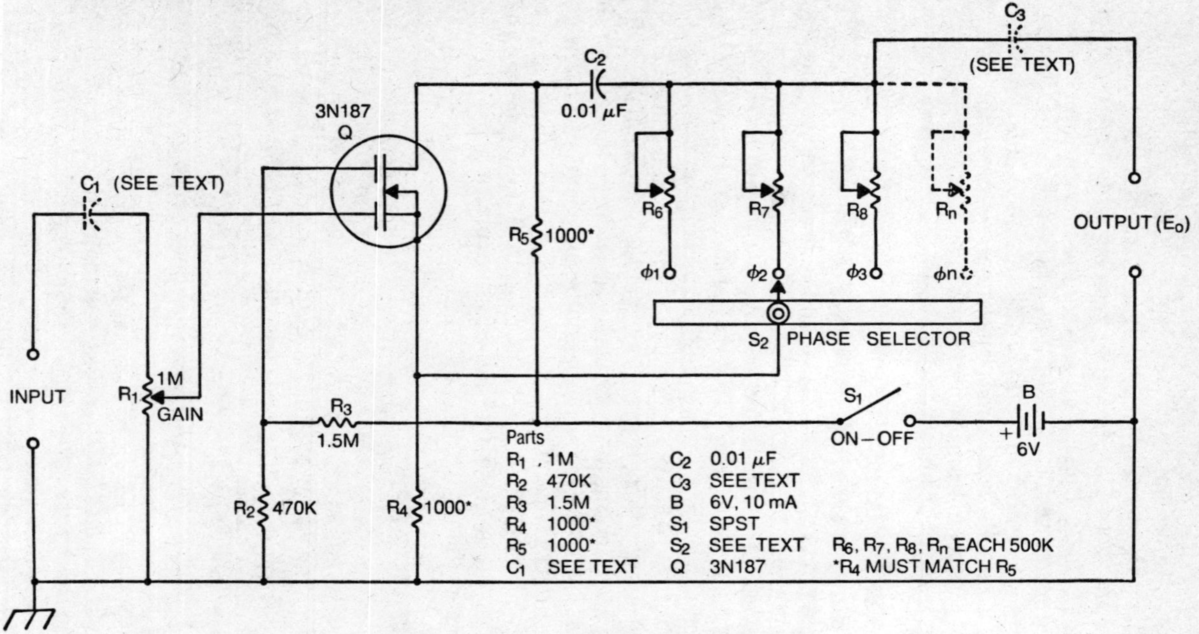

Fig. 4-14. Step-type phase shifter.

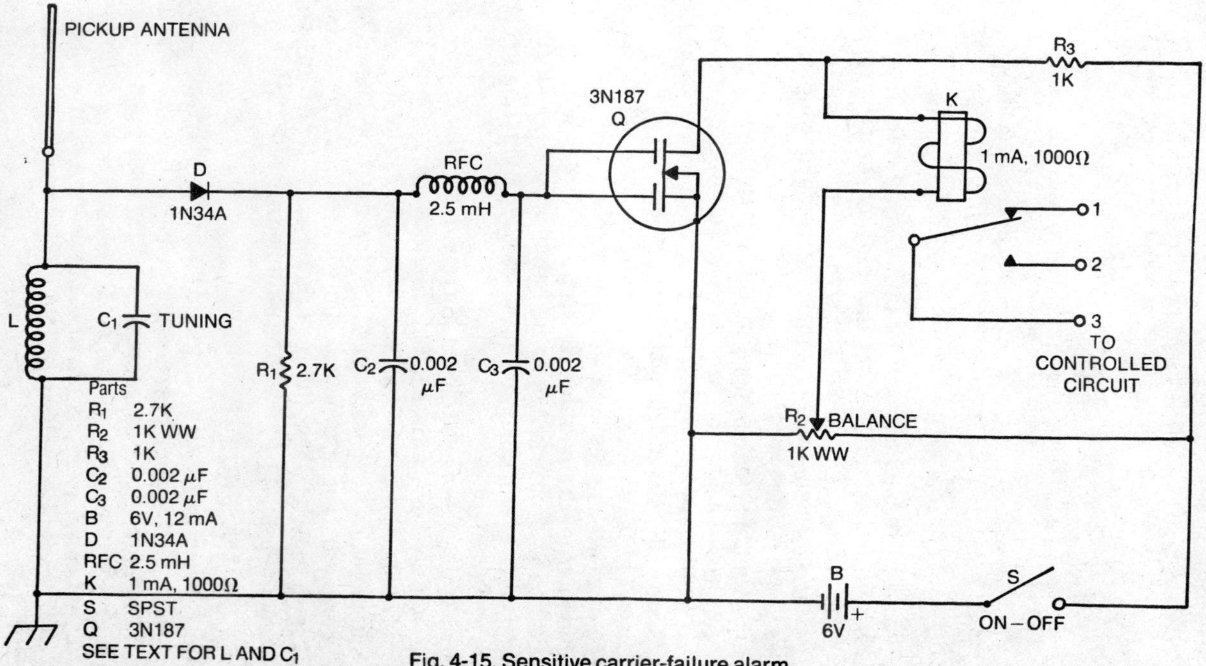

Fig. 4-15. Sensitive carrier-failure alarm.

Table 4-1. Coil Data for Sensitive Carrier-Failure Alarm.*

Coil A 440–1200 kHz	187 turns No. 32 enameled wire closewound on 1 in diameter form.
Coil B 1–3.5 MHZ	65 turns No. 32 enameled wire closewound on ½ in. diameter form.
Coil C 3.4–9 MHz	27 turns No. 26 enameled wire closewound on ½ in diameter form.
Coil D 8–20 MHz	10 turns No. 22 enameled wire on ½ in. diameter form. Space to winding length of ½ in.
Coil E 18–30 MHz	5½ turns No. 22 enameled wire on ½ in. diameter form. Space to winding length of ½ in.

*(based on tuning capacitor C_1 = 365 pF variable.)

standard, manufactured, transmitter-type coils to go with the chosen capacitance, when the wide coverage provided by Table 4-1 is not desired. The signal is picked up by a small antenna, which can be a standard vertical whip or a short section of stiff vertical wire.

When a signal is tuned in, diode *D* delivers a positive DC voltage to the gates of the MOSFET *Q*. The MOSFET amplifies the DC and drives the relay. The relay is a 1 mA, 1000Ω device (Sigma 5F-1000 or equivalent) operated in a 4-arm bridge circuit in the drain section. The bridge—consisting of resistor R_3, the two "halves" of balance-control potentiometer R_2, and the internal drain-to-source resistance of the MOSFET—allows the static drain current to be balanced out of the relay (by adjustment of potentiometer R_2) before attempting to operate the instrument.

When the relay is closed, current drain from the 6V supply is approximately 12 mA, but this may vary somewhat with individual MOSFETs. Use output terminals *1* and *3* when the circuit of the external alarm device must be closed by the relay actuation (example, a lamp or bell that is switched on); use *2* and *3* when the external circuit must be opened (example, a lamp that is to be switched off).

LOGIC AND CIRCUIT

Figure 4-16 shows a 2-stage AND circuit which performs the logic AND operation in control or counting circuits. The use of MOSFETs in this arrangement provides high input resistance and accordingly low drain on the switching source. The arrangement is essentially that of two DC amplifiers having a common source resistor, R_5. Output is taken from across this resistor.

The gates of each MOSFET receive a −1.5V switching signal. Normally, output E_o = +2V. If either one of the gates receives its signal, but the other gate does not, E_o is reduced

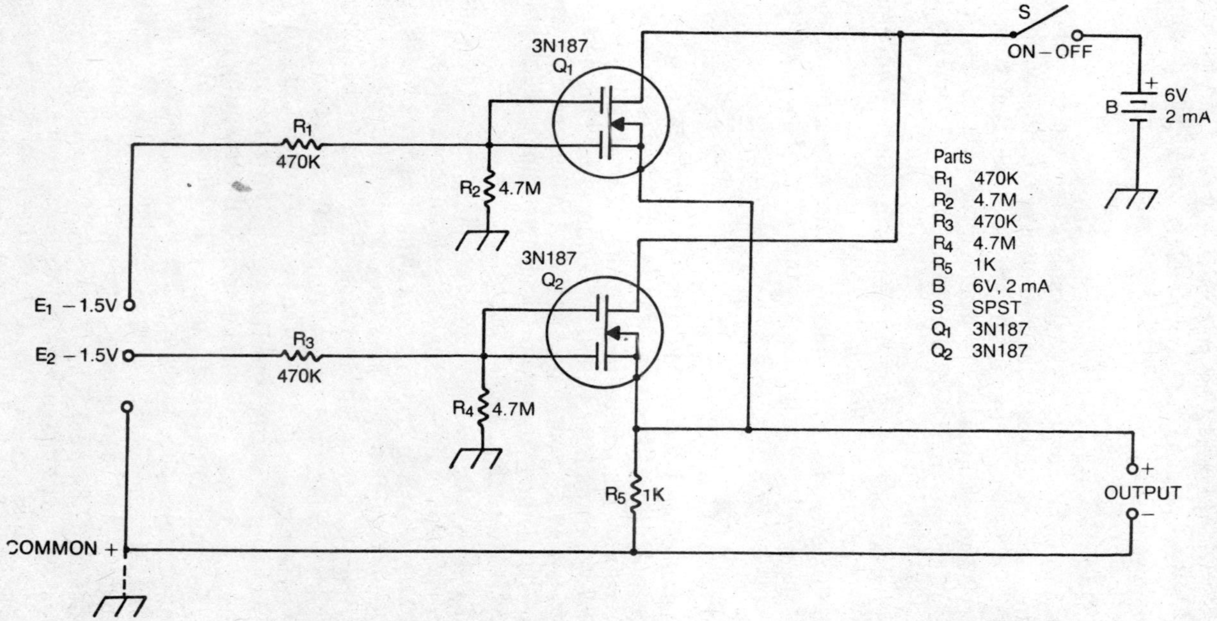

Fig. 4-16. Logic AND circuit.

slightly because of reduced drain current flowing through R_5. If both gates receive their −1.5V signals at the same time, however (that is, E_1 *and* E_2 are present), E_0 is reduced to zero.

When the circuit is performing the AND function, current drain from the 6V supply is approximately 2 mA, but this may vary somewhat with individual MOSFETs.

LOGIC OR CIRCUIT

Figure 4-17 shows a single-stage OR circuit which performs the logic OR operation in control or counting circuits. The use of a MOSFET in this arrangement provides high input resistance and accordingly low drain on the switching-signal source. The arrangement is essentially a dual-input, single-stage DC amplifier.

Each gate of the MOSFET receives a +1.5V switching signal. Initially, output voltage E_o = 1V. If either E_1 or E_2 is applied, however, E_o drops to 0.25V. Thus, either one of the input signals will perform the operation (E_1 *or* E_2 will switch the circuit).

DC SIGNAL INVERTER

In some control or counting circuits, a DC signal or trigger must be changed in polarity as well as amplified. Figure 4-18 shows one type of circuit for accomplishing this. This is seen to be a simple DC amplifier with a conventional bucking section (B_2–S_2–R_3) in its output. Use of a MOSFET in this arrangement provides high input impedance and accordingly negligible loading of the switching-signal source.

When the input switching signal is applied, the gates of the MOSFET become positive with respect to the source. This causes an increase in drain current, which reduces the drain voltage. Initially, however, the static output voltage has been balanced out by adjustment of potentiometer R_3; so the output voltage actually increases, but in the negative direction.

Initially, the circuit is balanced (with zero input signal) by adjusting balance control potentiometer R_3 for zero voltage at the output terminals. The circuit then will remain balanced unless the setting of R_3 is later disturbed. When a +1.5V signal subsequently is applied to the input terminals, the voltage at the output terminals becomes approximately −2.8V. Thus, the input voltage has been inverted and, at the same time, has been amplified 1.87 times.

When the circuit is in full operation, current drain from battery B_1 is approximately 10 mA, and that from battery B_2 is approximately 1.2 mA. The currents may be somewhat different from these values with individual MOSFETs.

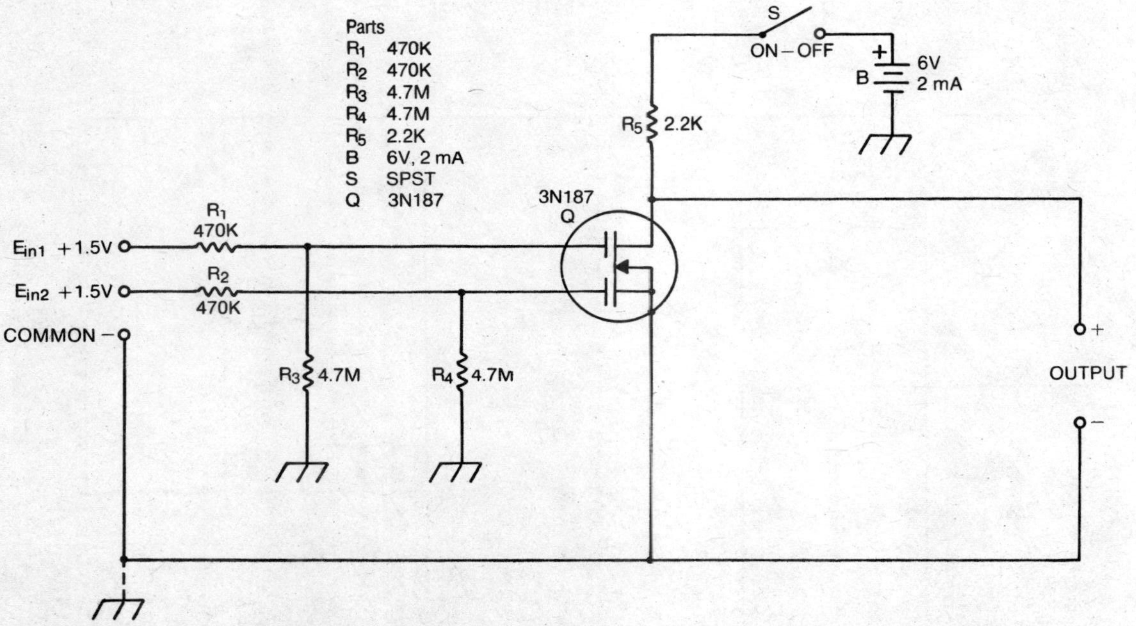

Fig. 4-17. Logic OR circuit.

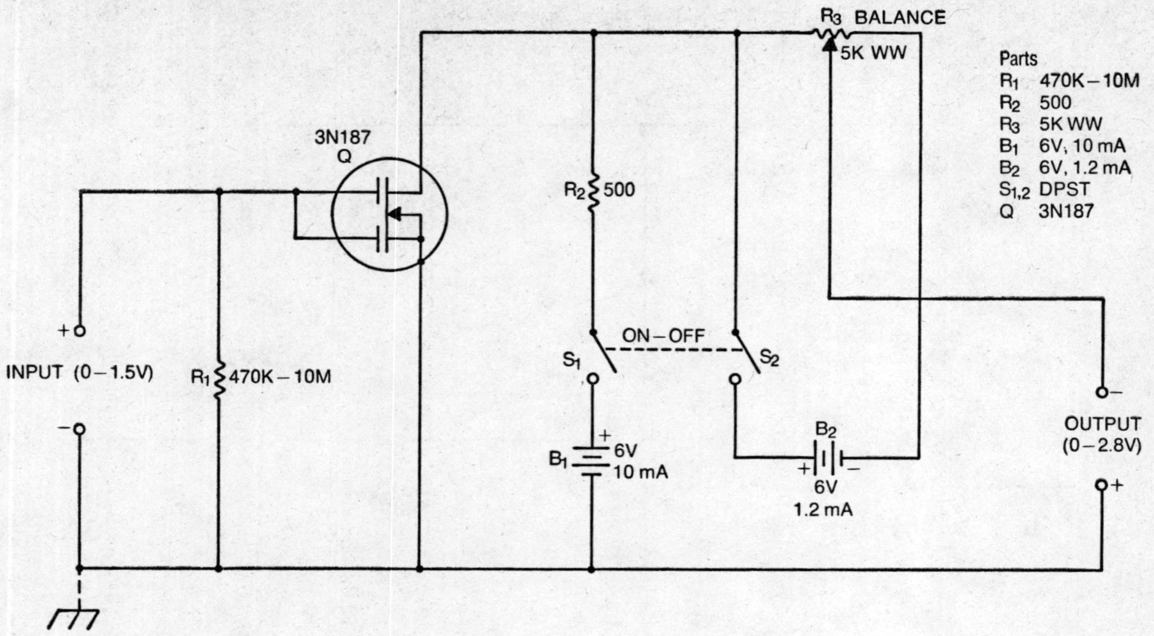

Fig. 4-18. Dc signal inverter.

DC IMPEDANCE CONVERTER

Figure 4-19 shows the circuit of a device that presents a high resistance to a DC source and a low resistance to a load. This is essentially a DC source follower, and it is useful in all control and instrumentation applications in which as much as possible of a DC voltage must be transmitted from a high-resistance source to a low-resistance load. The input resistance is determined chiefly by the full resistance of sensitivity control potentiometer R_1; however, beyond 5M or so, problems may be encountered with stray pickup. The output resistance is closely that of the the source resistor (here, 500Ω).

Initially, a static DC voltage appears at the output that is the result of the voltage drop produced by drain current through source resistor R_2. But this voltage is balanced out by adjustment of potentiometer R_3. The circuit then remains balanced unless the setting of this potentiometer is later disturbed. When a +1.5V signal subsequently is presented to the input terminals, with R_1 set for maximum sensitivity, the voltage at output terminals swings to +0.9V.

When the circuit is in full operation, current drain from battery B_1 is approximately 0.45 mA, and that from battery B_2 is approximately 2.6 mA. The output load device should not have a resistance lower than 2000Ω.

DUAL-POLARITY OUTPUT ADAPTER

The device whose circuit is given in Fig. 4-20 delivers both positive and negative outputs simultaneously in response to a positive input signal. Such a device is serviceable in certain counting and control systems. There are numerous ways of achieving dual-polarity DC output, but this is a simple method providing high input resistance and a common ground.

This circuit is seen to be similar to the inverter shown in Fig. 4-18, but with outputs taken from both the drain and source electrodes. Each output contains a conventional battery-and-rheostat bucking circuit to balance out the static voltage due to drain current through drain resistor R_5 and source resistor R_2. After the circuit is initially balanced with these elements (adjustment of rheostats R_3 and R_4), application of a +1.5V signal at the input terminals results in a −2.7V signal at the negative DC output terminal and a +0.9V signal at the positive DC output terminal.

During full operation of the circuit, current drain from each of the batteries is approximately 3.2 mA, but this may vary somewhat with individual MOSFETs.

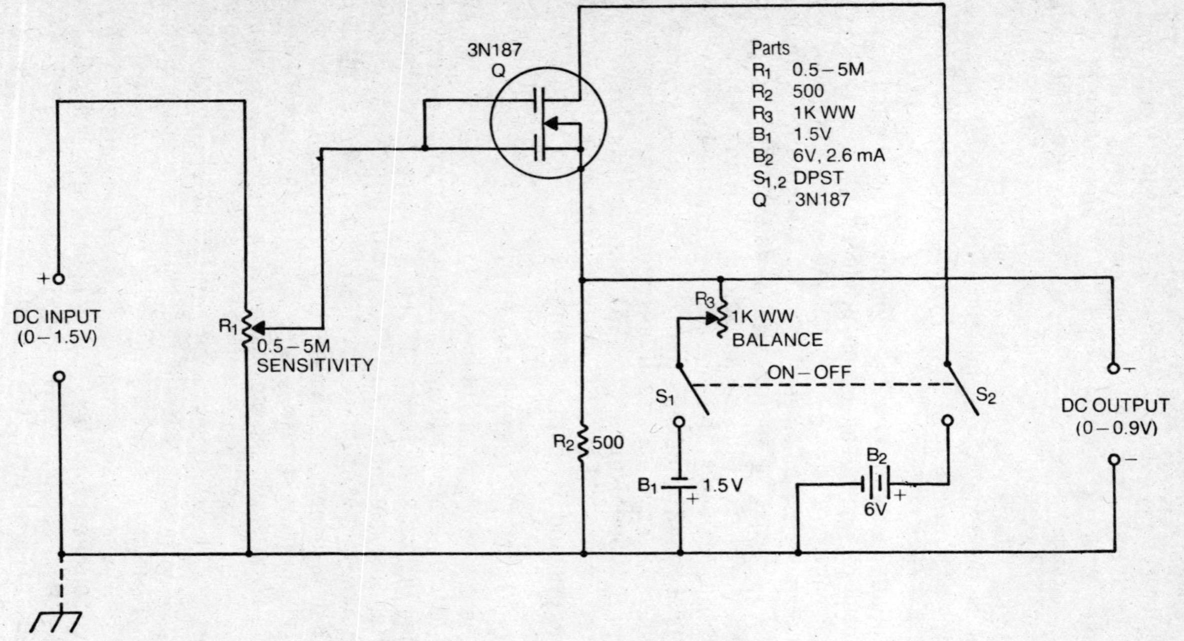

Fig. 4-19. Dc "impedance" converter.

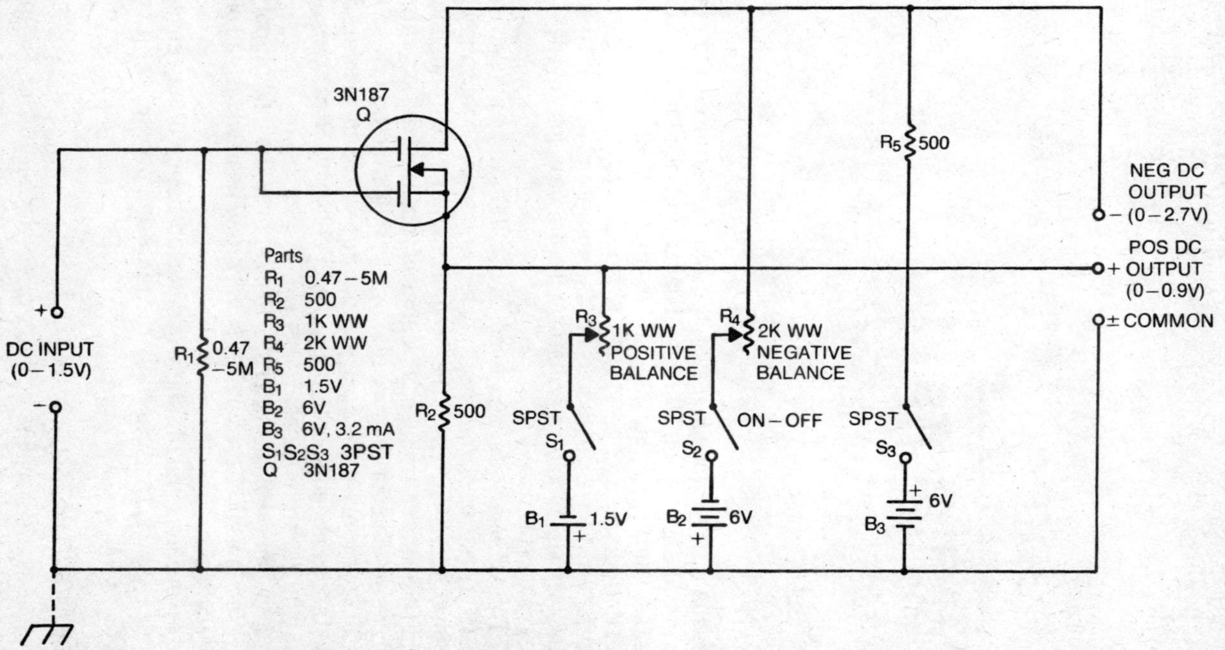

Fig. 4-20. Dual-polarity output adapter.

Chapter 5

Oscillators

The high input impedance of the MOSFET insures that this device, like the vacuum tube, will cause only negligible loading of the tuned circuit in an oscillator. Also, the high transconductance of the MOSFET usually results in vigorous oscillation. Moreover, some of the compromises which are necessary in bipolar oscillator circuits are not needed in MOSFET circuits.

This chapter offers 16 selected oscillator circuits: AF. RF. and IF. These arrangements may be used as is, or they may be readily modified by the experimenter for individual requirements of frequency, signal output, and DC operating voltage. In each of the circuits, unless indicated otherwise on the diagram or in the text, capacitances are in picofarads, resistances in ohms, and inductances in henrys. Resistors are ½W, and electrolytic capacitors are 25V.

For simplicity, batteries are shown for DC supply; however, a *well filtered* power supply may be used instead of a battery.

Before undertaking the wiring or operation of any circuit in this chapter, read carefully the hints and precautions given in Chapter 1.

TICKLER-TYPE AUDIO OSCILLATOR

One of the simplest oscillators employs a tickler coil for inductive feedback. In other words, a coupling transformer, phased correctly for positive feedback, is connected between the output and input circuits of a simple amplifier. This arrangement is also called an *Armstrong* oscillator. Figure 5-1 shows the circuit.

Here, transformer T is any convenient, miniature, transistor-type unit; its turns ratio (primary to secondary) should be 2:1. The primary winding of the transformer is tuned by capacitor C_1 to the desired operating frequency. This may be done experimentally, increasing C_1 to lower the frequency, and vice versa; or the selected winding may be measured for inductance, and the required capacitance calculated as follows:

$$C_1 = \frac{1}{39.5f^2L} \qquad \text{(Eq. 5-1)}$$

where C_1 is in farads, f in hertz, and L in henrys.
Thus, the primary winding of an Argonne AR-109 transformer has an inductance of 1.4H, and a capacitance of 0.018 μF will tune it to 1000 Hz. It must be remembered that the transformer windings have internal capacitance, and the upper frequency limit will be determined by this capacitance and the inductance of the winding.

Potentiometer R_1 permits close adjustment of the feedback voltage reaching gate 1 of the MOSFET. Too little voltage will not support oscillation, and too high a voltage will distort the signal waveform. This potentiometer is adjusted with the aid of an oscilloscope connected temporarily to the output terminals to monitor the waveform.

Gate 2 of the MOSFET is returned to the source. Gate 1 receives its negative bias from the voltage drop resulting from the flow of drain current through source resistor R_2. Current drain from the 6V supply is approximately 1 mA, but this may vary somewhat with individual MOSFETs. The open-circuit output voltage is approximately 2.7V RMS.

When setting up the circuit for the first time, if oscillation is not obtained, reverse either the primary or the secondary connections of transformer T. The color coding shown in Fig. 5-1 is correct for positive feedback.

HARTLEY AUDIO OSCILLATOR

In the Hartley oscillator circuit, a tapped tank coil is used; one part of this coil serves as the "input" winding, and the other part as the feedback winding. Figure 5-2 shows the circuit of a Hartley audio oscillator.

In this arrangement, the upper half of the centertapped transformer T serves as the input winding, applying a signal between gate 1 and source of the MOSFET; and the lower half serves as the feedback winding, through which the AC component of drain current flows from source to ground.

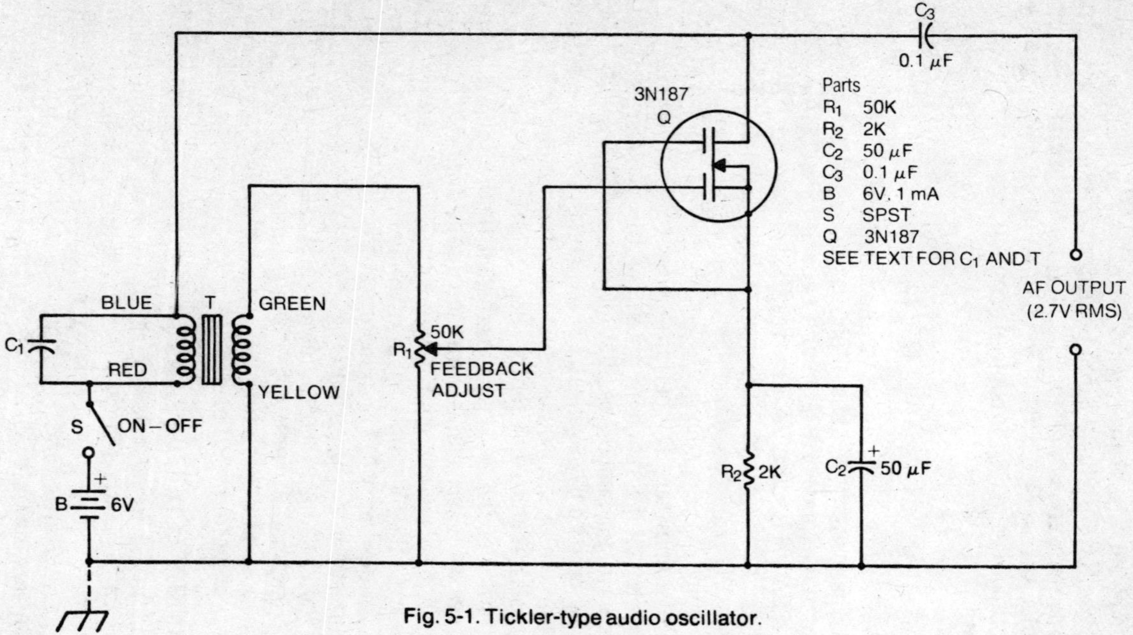

Fig. 5-1. Tickler-type audio oscillator.

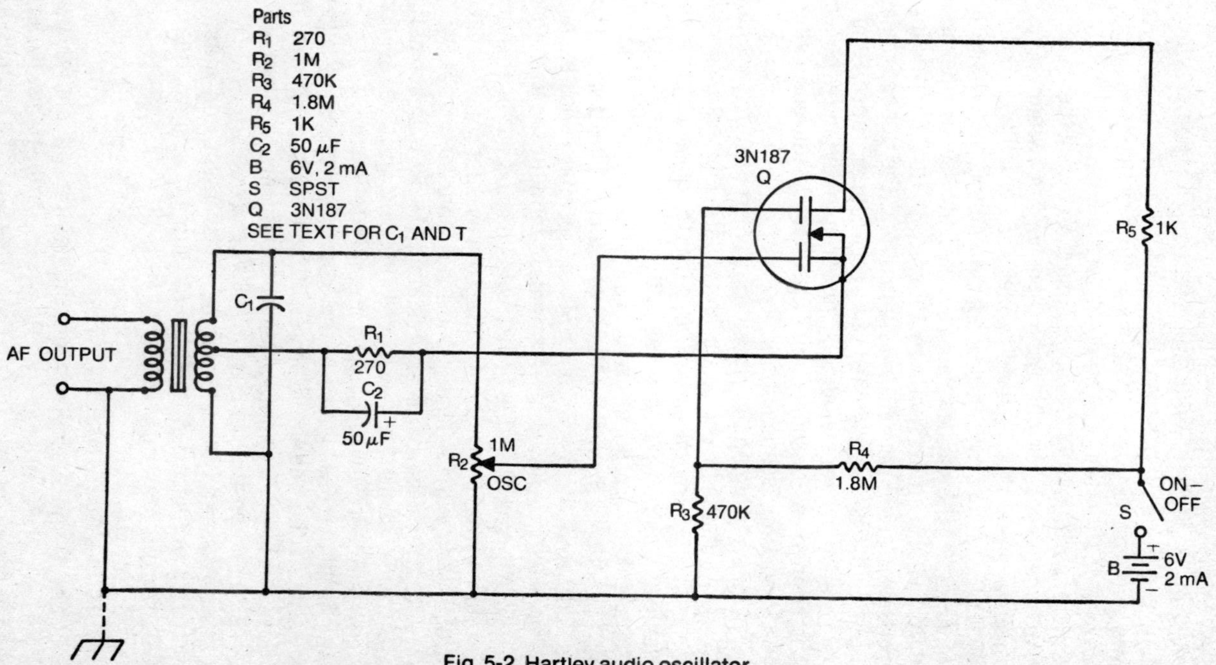

Fig. 5-2. Hartley audio oscillator.

Because the tapped winding is an *autotransformer*, the latter half inductively couples the feedback voltage into the former half.

Any convenient, miniature, transistor-type transformer having a 2:1 or 1:1 turns ratio and a center-tapped secondary winding can be used. The secondary winding is tuned by capacitor C_1 to the desired operating frequency. This may be done experimentally by increasing C_1 to lower the frequency, and vice versa; or the inductance of the winding may be measured and the required capacitance calculated by means of Eq. 5-1. Thus, with a secondary winding that shows 0.45H inductance, the capacitance required for 400 Hz operation is 0.352 μF. It must be remembered that the transformer windings have internal capacitance, and the upper frequency limit will be determined by this capacitance and the inductance of the winding. This frequency may be easily determined by disconnecting capacitor C_1 and observing the corresponding frequency.

When setting up the circuit for the first time, if oscillation is not obtained, check carefully all connections to the transformer. Potentiometer R_2 permits close adjustment of the feedback voltage reaching gate 1 of the MOSFET; too low a voltage will not produce oscillation, whereas too high a voltage will produce a distorted output waveform. This potentiometer is adjusted with the aid of an oscilloscope connected temporarily to the output terminals to monitor the waveform.

The output signal is inductively coupled out of the circuit by the primary winding of the transformer, and its actual amplitude will depend upon the characteristics of the transformer used. With an Argonne AR-109 transformer, the frequency is 400 Hz for $C_1 = 0.113$ μF, and the AF output voltage (open circuit) is approximately 1.4V RMS.

In this circuit, gate 1 of the MOSFET receives its negative bias from the voltage drop resulting from the flow of drain current through bypassed source resistor R_1; and gate 2 receives its positive bias from voltage divider R_3–R_4. R_4 may require some adjustment for best waveform and maximum output. Current drain from the 6V supply is approximately 2 mA, but this may vary somewhat with individual MOSFETs.

COLPITTS AUDIO OSCILLATOR

The Colpitts oscillator circuit has the advantage that it uses an untapped inductor. However, it does require a split tuning capacitor (see C_3 and C_4 in Fig. 5-3). The circuit shown here is conventional: One winding of transformer T serves as the frequency-determining inductor and is tuned by the

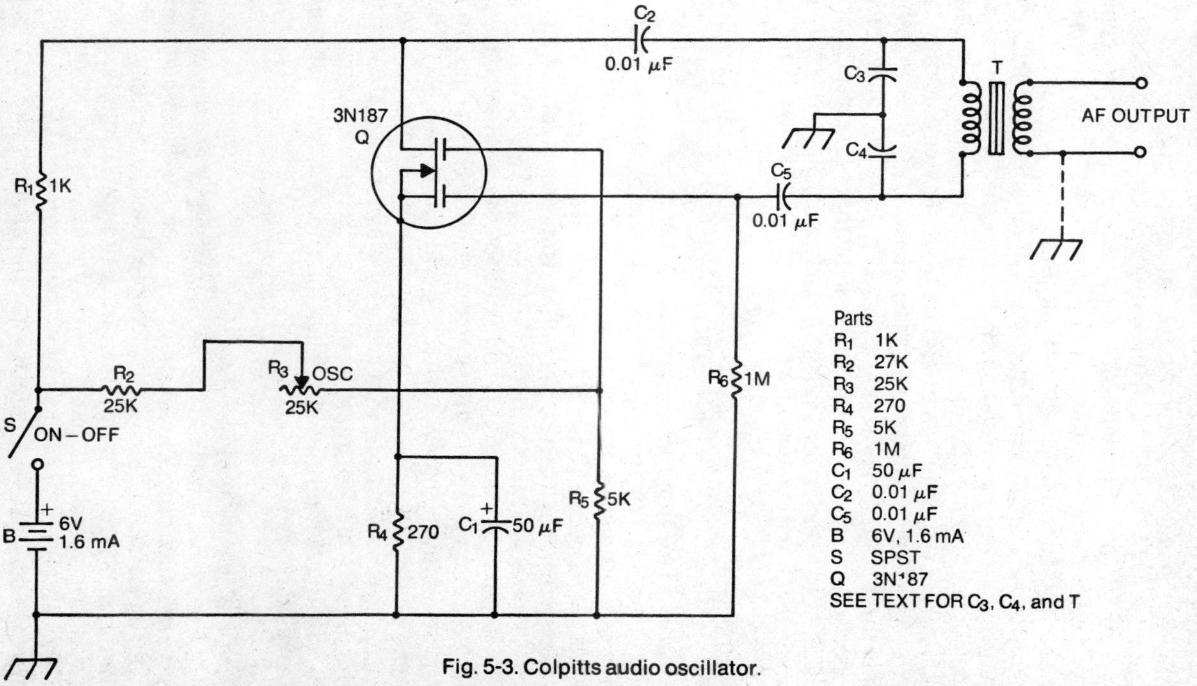

Parts	
R_1	1K
R_2	27K
R_3	25K
R_4	270
R_5	5K
R_6	1M
C_1	50 μF
C_2	0.01 μF
C_5	0.01 μF
B	6V, 1.6 mA
S	SPST
Q	3N187

SEE TEXT FOR C_3, C_4, and T

Fig. 5-3. Colpitts audio oscillator.

series-capacitor combination C_3–C_4 (identical capacitances). Capacitors C_2 and C_5 are blocking units only, serving to keep the transformer from short-circuiting the drain electrode of the MOSFET to the gate 1 electrode. Operation of the circuit is somewhat similar to that of the Hartley circuit (see preceding section), except that the tuning capacitance here is tapped instead of the transformer winding.

When C_3 and C_4 are identical, the equivalent capacitance tuning the primary winding of transformer T is equal to one-half of either C_3 or C_4. Any convenient, miniature, transistor-type transformer can be used. The circuit may be adjusted experimentally to the desired operating frequency by increasing C_3 and C_4 (in identical pairs) to lower the frequency, and vice versa. Or the inductance of the primary winding of transformer T may be measured, and the required value for C_3 and C_4 calculated as follows:

$$C_3 = C_4 = \frac{2}{39.5f^2L} \qquad \text{(Eq. 5-2)}$$

where C is in farads, f in hertz, and L in henrys.

Thus, the primary winding of an Argonne AR-109 transformer has an inductance of 1.4H, and a capacitance of 0.145 μF each for C_3 and C_4 will tune it to 500 Hz. With this transformer, the open-circuit output voltage is approximatelv 1V RMS.

In the Colpitts circuit, gate 1 of the MOSFET receives its negative bias from the voltage drop resulting from the drain current through source resistor R_4, and its positive bias from voltage divider R_2–R_3–R_5. By providing control of this voltage, rheostat R_3 in this voltage-divider network controls the strength of oscillation. Current drain from the 6V supply is approximately 1.6 mA, but this may vary somewhat with individual MOSFETs.

FRANKLIN AUDIO OSCILLATOR

Figure 5-4 shows the circuit of a Franklin oscillator. This arrangement consists essentially of a 2-stage audio amplifier with the output of the last stage coupled back to the input of the first stage through a capacitor (C_4) to provide the feedback necessary for oscillation. This oscillator has the advantage that a single, grounded, untapped tuned circuit (L–C_1) can be used, and that the signal waveform may be purified considerably by use of a high-Q inductor and close adjustment of oscillation-control potentiometer R_{10}.

The frequency is determined by inductance L and capacitance C_1. The best procedure in setting up this tuned

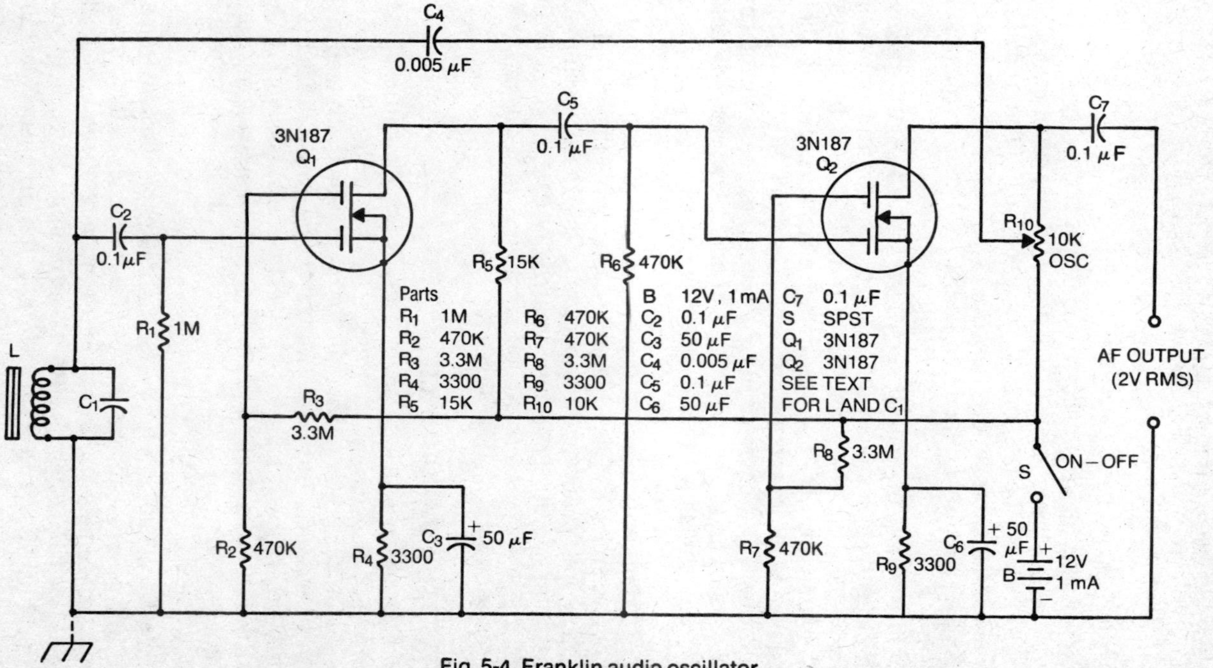

Fig. 5-4. Franklin audio oscillator.

circuit is to obtain a convenient inductor and then prune the capacitance. The following formula may be employed to determine the required value of C_1:

$$C_1 = \frac{1}{39.5f^2L} \qquad \text{(Eq. 5-3)}$$

where C_1 is in farads, f in hertz, and L in henrys.

In the author's circuit L is a high-Q, 5.4H inductor (United Transformer VI-C15), and C_1 is 0.0047 μF. The resulting frequency is approximately 1 kHz. (The VI-C15 is adjustable over a narrow range, and can be set for exact frequency, if desired.)

In Fig. 5-4, the gate-1 electrodes of the MOSFETs receive their negative bias from the voltage drop resulting from the flow of drain current through source resistors (R_4 for MOSFET Q_1, and R_9 for Q_2), and the gate-2 electrodes receive their positive bias from voltage dividers (R_2–R_3 for MOSFET Q_1, and R_7–R_8 for Q_2). For maximum AF output and best waveform, resistors R_3 and R_8 in the voltage-divider networks may need some adjustment with individual MOSFETs.

Oscillation control pot R_{10} must be set for the feedback amplitude that will afford the best sine-wave voltage at the output terminals. This should be done with the aid of an oscilloscope connected temporarily to those terminals. Maximum AF output is approximately 2V RMS. Current drain from the 12V source B is approximately 1 mA, but this may vary somewhat with individual MOSFETs.

PHASE-SHIFT AUDIO OSCILLATOR

Fgure 5-5 shows the circuit of an RC-tuned audio oscillator employing a 3-leg RC network (C_1–C_2–C_3–R_3–R_4–R_5) to obtain the necessary phase shift for oscillation in the feedback path between the drain of MOSFET Q and gate 1. (Each leg shifts the phase 60°, for a total ot 180° for the full network.) The phase-shift oscillator is noted for its compactness and excellent waveform and is often used in vacuum-tube equipment.

In the phase-shift network, all three capacitors are equal, and all three resistors are equal. The frequency of the network (and accordingly the oscillation frequency of the circuit) is $f = 1/(15.4RC)$, where f is in hertz, R in ohms, and C in farads. From this it is easily seen that the frequency of the circuit in Fig. 5-5 (where $C_1 = C_2 = C_3 = 0.1$ μF, and $R_3 = R_4 = R_5 = 1000\Omega$) is 649 Hz. In general, it will be easiest to select available equal capacitors and then to prune the

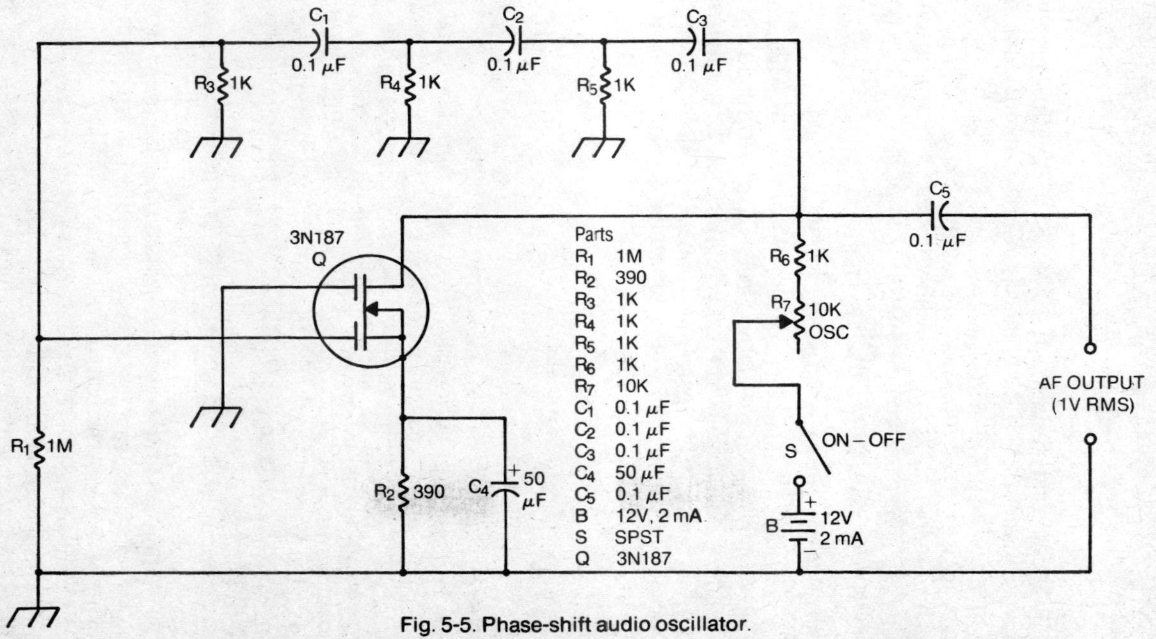

Fig. 5-5. Phase-shift audio oscillator.

resistances to required values. This may be done with the aid of the following equation:

$$R = \frac{1}{15.4fC} \qquad \text{(Eq. 5-4)}$$

For best results, the MOSFET should have high transconductance (i.e., higher than the typical value for its type). Adjustment of the oscillation control R_7 allows the circuit to be set for fastest starting. Signal voltage at the output terminals is approximatey 1V RMS. Current drain from the 12V source is approximately 2 mA, but this may vary somewhat with individual MOSFETs.

WIEN-BRIDGE AUDIO OSCILLATOR

Figure 5-6 shows a MOSFET version of the popular Wien-bridge audio oscillator circuit. This *RC*-tuned circuit is noted for its excellent sine-wave output and its comparatively simple tuning. The operating frequency is determined by the *RC* network C_1–C_2–R_1–R_3, in which $C_1 = C_2$ and $R_1 = R_3$, by the formula

$$f = \frac{1}{6.28R_1C_2} \qquad \text{(Eq. 5-5)}$$

where f is in hertz, R_1 in ohms, and C_2 in farads.

By varying either the resistance *or* the capacitance simultaneously, you can obtain a variable-frequency oscillator of considerable usefulness. In fact, many audio signal generators employ the Wien-bridge circuit.

The basis of the circuit is a 2-stage *RC*-coupled amplifier with the Wien bridge in the feedback loop between the drain of MOSFET Q_2 and gate 1 of Q_1. The bridge capacitors and resistors in Fig. 5-6 give an operating frequency of 1061 Hz. The resistors may be pruned to 31,847Ω each to give 1000 Hz operation. The oscillation-adjust rheostat R_7 must be set for oscillation, and the waveform-adjust potentiometer R_9 set for the best sine waveform, as viewed with an oscilloscope connected temporarily to the output terminals.

Each of the MOSFETs receives its negative gate-1 bias from the voltage drop resulting from the flow of drain current through a source resistor (R_5 for Q_1, and R_{11} for Q_2), and its positive gate-2 bias from a voltage divider (R_4–R_6 for Q_1, R_8–R_{10} for Q_2). In these dividers, resistances R_6 and R_8 may need some adjustment with individual MOSFETs for best waveform.

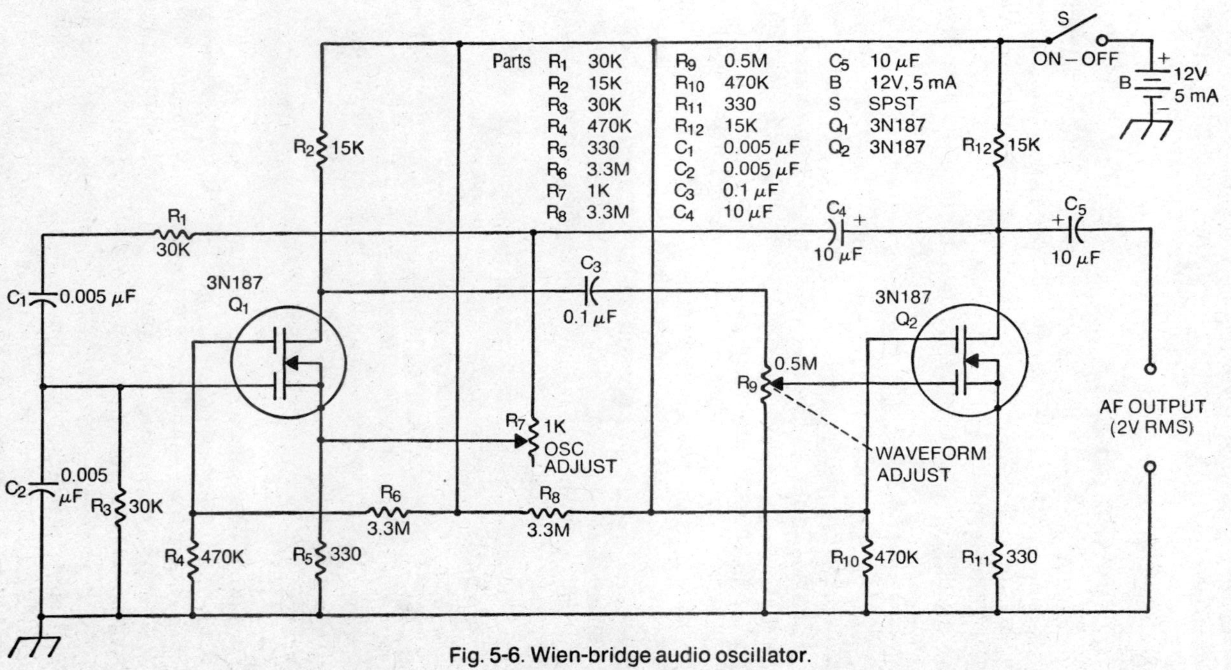

Fig. 5-6. Wien-bridge audio oscillator.

The output signal amplitude is approximately 2V RMS. Somewhat lower output voltage, at low impedance, may be obtained across resistor R_{11}. If desired, resistor R_{11} may be replaced with a potentiometer for output voltage adjustment. Current drain from the 12V supply is approximately 5 mA, but this may vary somewhat with individual MOSFETs.

VILLARD AUDIO OSCILLATOR

Figure 5-7 shows the circuit of another *RC*-tuned audio oscillator having excellent sine-wave output. This circuit covers the range 200 Hz to 10 kHz in one rotation of the dual potentiometer R_5–R_{11}.

In this arrangement, tuning is accomplished with two phase-shift networks, C_2–R_5 and C_3–R_{11}, each of which provides a total shift of 90°. The output of the 3-stage amplifier Q_1–Q_2–Q_3 is fed back to the input stage (via capacitor C_1) in the correct phase for positive feedback (and, accordingly, oscillation), after having undergone phase rotation in the Q_4 auxiliary stage. (See the section *Continuously Variable Phase Shifter*, in Chapter 4, for further discussion.)

Oscillation control potentiometer R_{17} is set for oscillation at any setting of the frequency control R_5–R_{11}. Potentiometer R_{17} may then need to be reset for restoration of oscillation if R_5–R_{11} is moved to a distant setting. (When oscillation is obtained at the low-frequency end of the frequency control, it will disappear at the high-frequency end, and vice versa, unless R_{17} is reset.)

The output signal amplitude at the AF output terminals is approximately 2.5V RMS. A gain control potentiometer (e.g., 10K) may be connected across these terminals if an adjustable voltage is desired. The frequency control potentiometer may be provided with a dial calibrated to read directly in hertz. Current drain from the 6V supply is approximately 40 mA, but this may vary somewhat with individual MOSFETs.

MULTIVIBRATOR

Figure 5-8 shows the circuit of a conventional drain-coupled multivibrator. This circuit is equivalent to the plate-coupled vacuum-tube multivibrator. With the circuit constants given in Fig. 5-8, the output is an approximately square waveform of 4V peak-to-peak amplitude.

As in the corresponding tube circuit, the operating frequency depends upon the resistance and capacitance coupling components. Thus, $C_2 = C_3$, $R_1 = R_4$, and $f = 1/(2\pi R_1 C_2)$, where f is in hertz, C_2 in farads, and R_1 in ohms. The frequency may easily be changed to another

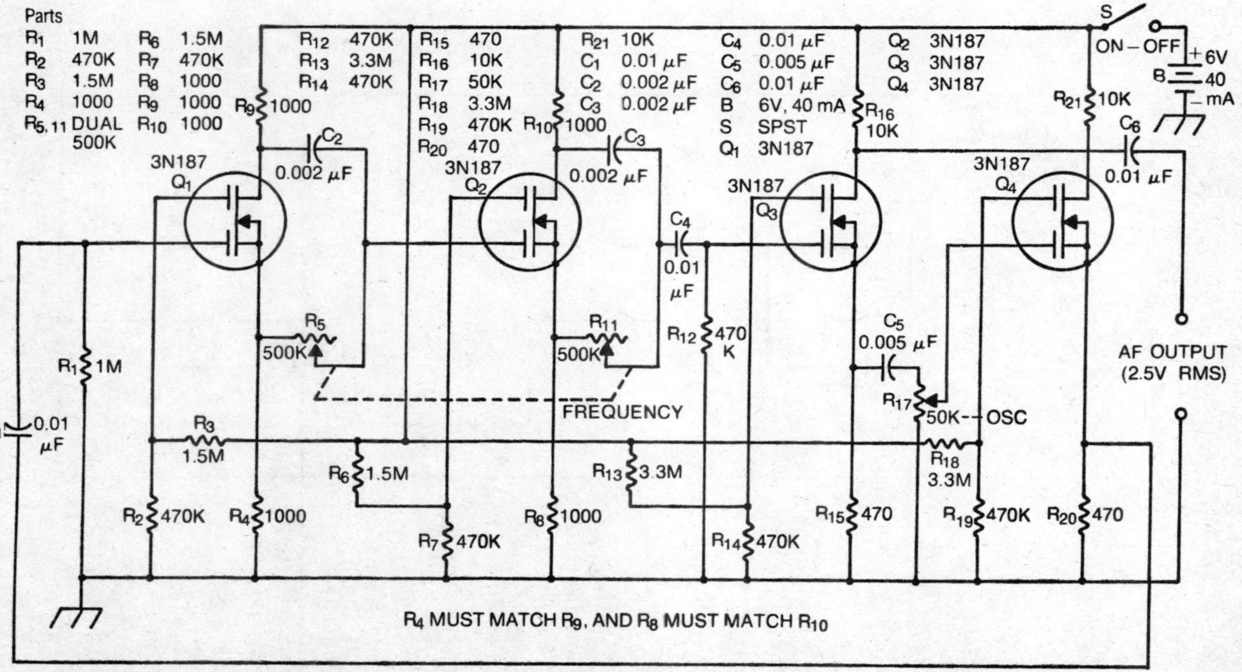

Fig. 5-7. Villard audio oscillator.

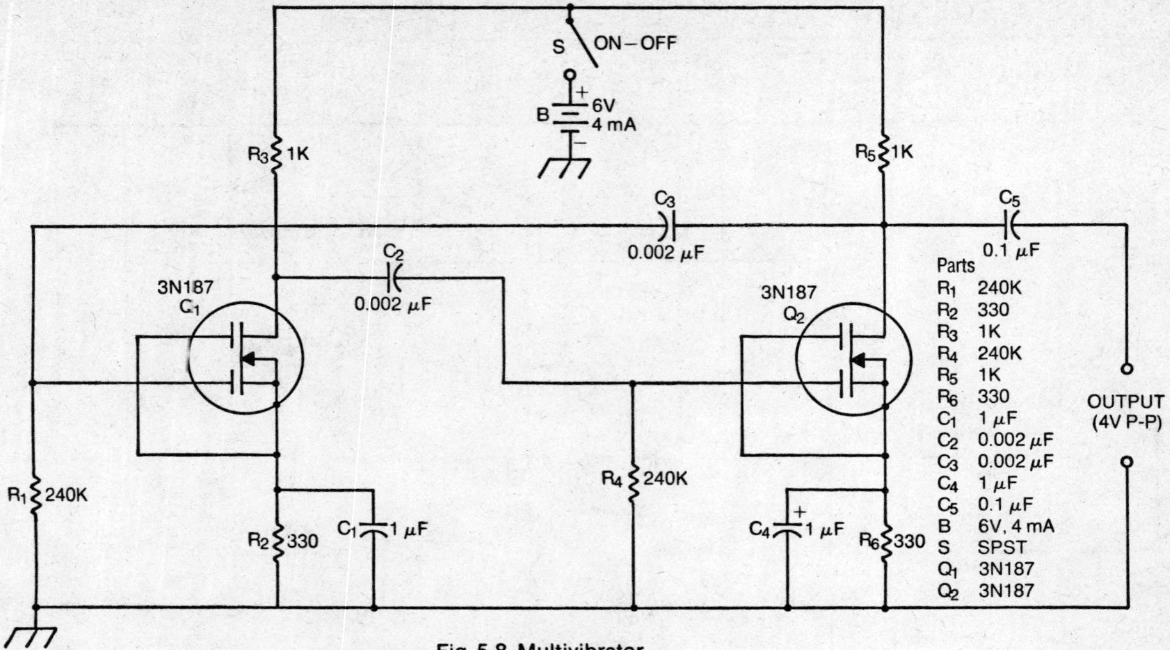

Fig. 5-8. Multivibrator.

desired value by changing the resistances and capacitances while preserving the relationships given above.

The multivibrator is essentially a positive-feedback amplifier in which the drain of each stage feeds gate 1 of the other stage. The gate-2 electrodes are returned to the sources. The output wave will have best symmetry when the two MOSFETs are closely matched; or, lacking this, when either R_2 or R_6 is adjusted for matching of operating points.

Current drain from the 6V supply is approximately 4 mA, but this may vary somewhat with individual MOSFETs.

SUN-POWERED AUDIO OSCILLATOR

Figure 5-9 shows the circuit of a tickler-type audio oscillator of the same type as that described earlier, but with the battery replaced by two silicon solar cells PC_1 and PC_2. In bright sunlight, these two cells (International Rectifier S7M-C) deliver a total output of 6V and will adequately power the oscillator.

The output signal amplitude (open-circuit) is approximately 2.7V RMS; if transformer T, recommended in the earlier circuit at the beginning of the chapter, is employed with a C_1 capacitance of 0.018 μF, the frequency is approximately 1000 Hz.

See the earlier section for a description of the amplifier circuit alone.

SELF-EXCITED RF TICKLER-COIL OSCILLATOR

In Fig. 5-10, an Armstrong-type radio-frequency oscillator, oscillation is obtained with a tickler coil, L_1, which feeds energy from the output of the circuit into the input tuned circuit L_2–C_2. The operating frequency is determined by the values of L_2 and C_2 according to

$$f = \frac{1}{6.28\sqrt{L_2C_2}} \qquad \text{(Eq. 5-6)}$$

where f is in hertz, C_2 in farads, and L_2 in henrys.

It is helpful to know also that, from rewriting Eq. 5-6, $L_2 = 1/39.5\ f^2C_2$ and $C_2 = 1/39.5\ f^2L_2$. When C_2 is a 365 pF variable capacitor, coils may be wound as shown in Table 5-1 to cover the frequency range of 440 kHz to 30 MHz in five bands.

In this circuit, MOSFET Q receives its negative gate-1 bias from the voltage resulting from the flow of drain current through source resistor R_3, and its positive gate-2 bias from voltage divider R_1–R_2. Resistance R_1 of this divider may

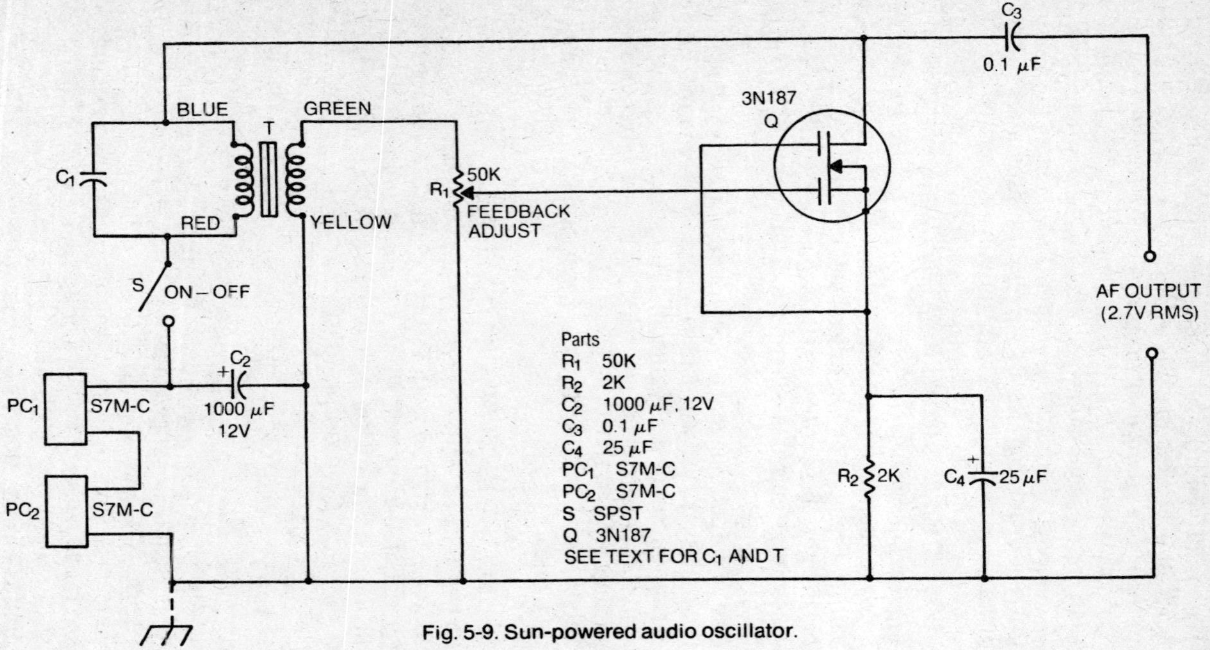

Fig. 5-9. Sun-powered audio oscillator.

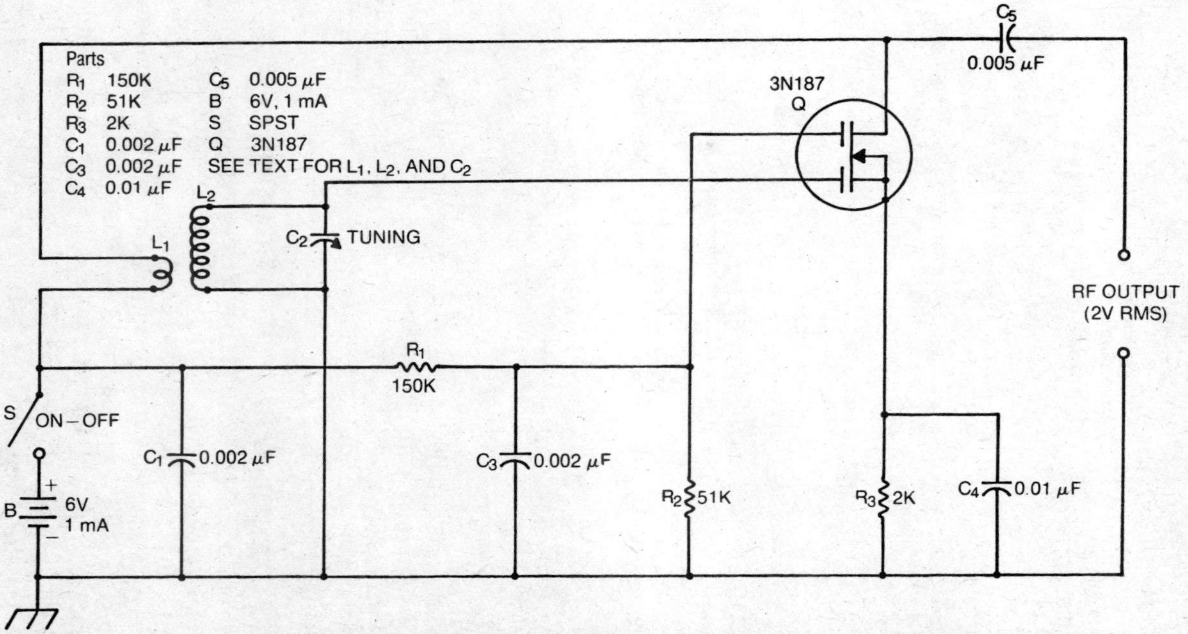

Fig. 5-10. Self-excited RF oscillator (tickler type).

Table 5-1. Coil-Winding Data for Tickler-Type RF Oscillator.*

Band	Coil	Winding
Band A 440–1200 kHz	L_1	45 turns No. 32 enameled wire closewound on same form as L_2. Space 1/16 in. from top of L_2.
	L_2	187 turns No. 32 enameled wire closewound on 1 in. diameter form.
Band B 1–3.5 MHz	L_1	15 turns No. 32 enameled wire closewound on same form as L_2. Space 1/16 in. from top of L_2.
	L_2	65 turns No. 32 enameled wire closewound on 1 in. diameter form.
Band C 3.4–9 MHz	L_1	8 turns No. 26 enameled wire closewound on same form as L_2. Space 1/16 in from top of L_2.
	L_2	27 turns No. 26 enameled wire closewound on 1 in. diameter form.
Band D 8–20 MHz	L_1	4 turns No. 22 enameled wire closewound on same form as L_2. Space 1/16 in. from top of L_2.
	L_2	10 turns No. 22 enameled wire closewound on 1 in. diameter form.
Band E 18–30 MHz	L_1	4 turns No. 22 enameled wire airwound ½ in. in diameter. Mount 1/16 in. from top of L_2.
	L_2	5½ turns No. 22 enameled wire airwound ½ in. in diameter. Space to winding length of ½ in.

*(Tuning capacitance C_2 = 365 pF.)

require some adjustment with an individual MOSFET for maximum RF output. The output signal amplitude is approximately 2V RMS. Current drain from the 6V supply is approximately 1 mA, but this may vary somewhat with individual MOSFETs.

SELF-EXCITED RF HARTLEY OSCILLATOR

The Hartley oscillator circuit requires only one tuned-circuit inductor; a tap on this inductor allows one portion to act as a feedback (tickler) coil, and the other portion as the tuned-circuit (tank) coil. The operating frequency of such a circuit (see Fig. 5-11) is determined by the whole inductance L and the capacitance of the tuning capacitor C_1, according to Eq. 5-6.

For the circuit in Fig. 5-11, the reader may employ manufactured coils for a desired frequency range and supply the recommended C_1 value for these coils; or he may employ a 365 pF variable capacitor for C_1 and wind his own coils

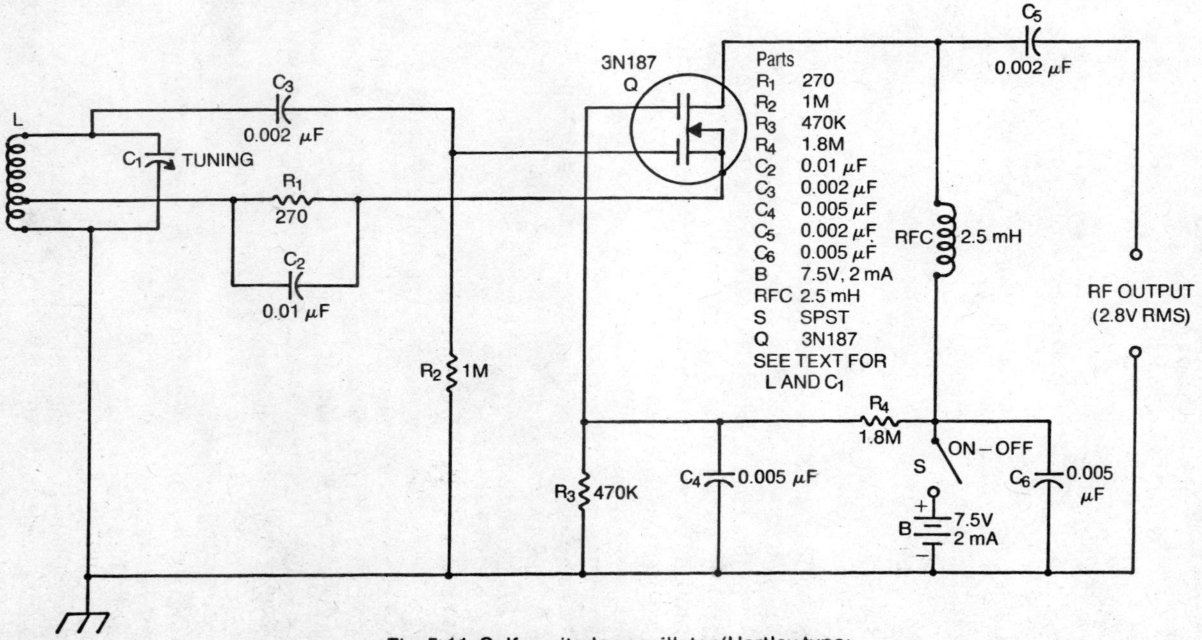

Fig. 5-11. Self-excited RF oscillator (Hartley type).

Table 5-2. Coil-Winding Data for Hartley RF Oscillator.*

Band A 440–1200 kHz	187 turns No. 32 enameled wire closewound on 1 in. diameter form. Tap 45th turn from ground end.
Band B 1–3.5 MHz	65 turns No. 32 enameled wire closewound on 1 in. diameter form. Tap 16th turn from ground end.
Band C 3.4–9 MHz	27 turns No. 26 enameled wire closewound on 1 in. diameter form. Tap seventh turn from ground end.
Band D 8–20 MHz	10 turns No. 22 enameled wire closewound on 1 in. diameter form. Tap second turn from ground end.
Band E 18–30 MHz	5½ turns No. 22 enameled wire airwound ½ in. in diameter and spaced to winding length of ½ in. Tap second turn from ground end.

*(Tuning capacitance C_1 365 pF.)

according to instructions given in Table 5-2, to give a frequency coverage of 440 kHz to 30 MHz in five bands.

In this circuit, MOSFET Q receives its negative gate-1 bias from the voltage drop resulting from the flow of drain current through RF-bypassed source resistor R_1, and its positive gate-2 bias from voltage divider R_3–R_4. Resistance R_4 in this divider may require some adjustment with an individual MOSFET for maximum RF output voltage.

The amplitude of the RF output voltage is approximately 2.8V RMS. Current drain from the 7.5V supply is approximately 2 mA, but this may vary somewhat with individual MOSFETs.

SELF-EXCITED RF COLPITTS OSCILLATOR

Figure 5-12 shows the circuit of a Colpitts self-excited radio-frequency oscillator. This circuit, like the Colpitts audio oscillator circuit described earlier, employs an untapped inductor (L_1), but a tapped capacitor leg (C_6–C_7).

A dual variable capacitor (C_6–C_7) tunes inductor L_1 to the desired frequency. Radio-frequency energy is coupled out of the circuit by means of low-impedance coil L_2. Refer back to the section *Colpitts Audio Oscillator* and Eq. 5-2 for instructions for calculating the frequency of the Colpitts circuit in terms of inductance L_1 and the equivalent capacitance of C_6 and C_7 in series. Table 5-3 lists commercial RF coils which may be used with a dual 100 pF variable capacitor to cover the four frequency bands shown, extending from 1.1 MHz to 25 MHz. These coils are slug-tuned and can be preset for the exact band limits.

In this circuit, MOSFET Q receives its negative gate-1 bias from the voltage drop resulting from the flow of drain current

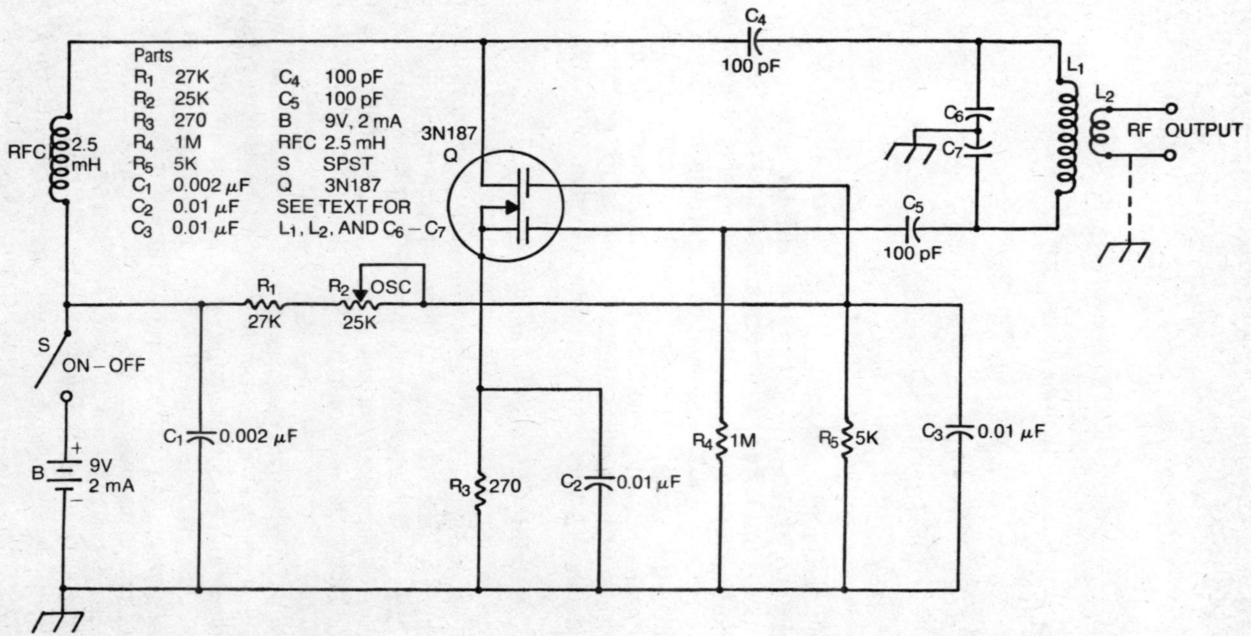

Fig. 5-12. Self-excited RF oscillator (Colpitts type)

Table 5-3. Coil Data for Colpitts RF Oscillator.*

Band	Coil	Data
Band A 1.1–2.5 MHz	L_1	Miller No. 20A474RBI (365–564 μH).
	L_2	5 turns No. 24 enameled wire wound close to L_1.
Band B 2.5–5 MHz	L_1	Miller No. 20A104RBI (88.5–120 μH).
	L_2	5 turns No. 24 enameled wire wound close to L_1.
Band C 5–11.5 MHz	L_1	Miller No. 20A225RBI (16.2–26.4 μH).
	L_2	2 turns No. 24 enameled wire wound close to L_1.
Band D 10–25 MHz	L_1	Miller No. 20A156RBI (1.08–1.80 μH).
	L_2	1 turn No. 24 enameled wire wound close to L_1.

*(Tuning capacitance C_6–C_7 = Dual 100 pF)

through source resistor R_3, and its positive gate-2 bias from voltage divider R_1–R_2–R_5. In this divider network, the oscillation control is set for maximum RF output voltage.

The RF output voltage will be of the order of 0.25V RMS. Higher voltage may be obtained by discarding L_2 and taking the output (through a 100 pF fixed capacitor) from the junction of the radio-frequency choke and the drain electrode of the MOSFET. Current drain from the 9V supply is approximately 2 mA, but this may vary somewhat with individual MOSFETs.

STANDARD CRYSTAL OSCILLATOR

Figure 5-13 shows the circuit of a conventional crystal oscillator in which the MOSFET replaces the vacuum tube or bipolar transistor. The only strange component is apt to be capacitor C_1, connected externally between drain and gate 1. This capacitance is required for feedback coupling, since in most MOSFETs the internal capacitance is too small to sustain oscillation.

In all other respects, the circuit is conventional. The resonant circuit L–C_3 is tuned to resonate at the crystal frequency, and RF output voltage is taken from the top of the tuned circuit through capacitor C_4. The required L and C_3 combination may be determined with the aid of Eq. 5-6 and its accompanying discussion. Or, if a 100 pF variable capacitor is employed for C_3, data will be found in Table 5-4 for coils covering the frequency range 680 kHz to 34 MHz in five bands.

The RF output voltage is approximately 3V RMS. Current drain from the 12V supply is approximately 2 mA, but this may vary somewhat with individual MOSFETs. The MOSFET receives its negative gate-1 bias from the voltage drop resulting from

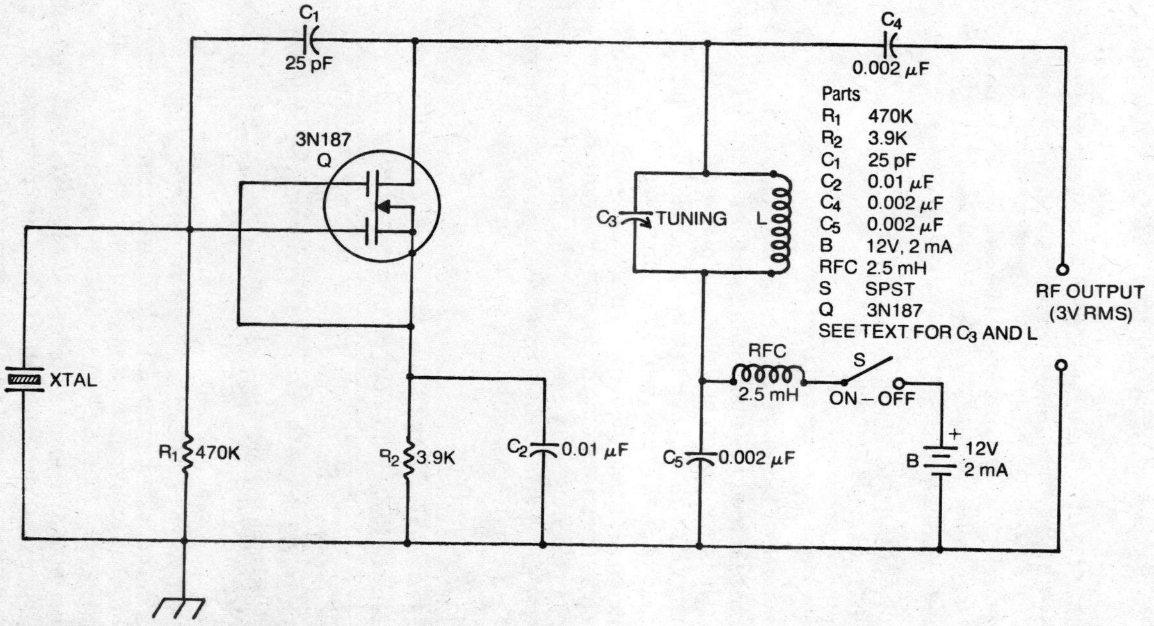

Fig. 5-13. Standard crystal oscillator.

Table 5-4. Coil-Winding Data for Standard Crystal Oscillator.*

Band	Winding
Band A 680 kHz – 1.5 MHz	181 turns No. 36 enameled wire closewound on 1 in. diameter form.
Band B 1.8 – 4 MHz	57 turns No. 32 enameled wire closewound on 1 in. diameter form.
Band C 3.8 – 8.6 MHz	25 turns No. 26 enameled wire on 1 in. diameter form. Space to winding length of ½ in.
Band D 8 – 18 MHz	12 turns No. 22 enameled wire on 1 in. diameter form. Space to winding length of ½ in.
Band E 15 – 34 MHz	5½ turns No. 22 enameled wire on 1 in. diam eter form. Space to winding length of ¼ in.

*(Tuning capacitance C_3 = 100 pF.)

the flow of drain current through source resistor R_2. Gate 2 is returned externally to the source.

PIERCE CRYSTAL OSCILLATOR

A MOSFET Pierce crystal oscillator is shown in Fig. 5-14. The Pierce circuit has the advantage that it requires no tuned circuit. It is not useful, however, for overtone crystals, since these crystals will oscillate in this circuit at their fundamental frequency rather than at the desired harmonic.

The open-circuit RF output voltage is approximately 3.2V RMS. Current drain from the 6V supply is approximately 1.6 mA, but this may vary somewhat with individual MOSFETs. The MOSFET receives its negative gate-1 bias from the voltage drop resulting from the flow of drain current through source resistor R_2.

SUN-POWERED RF OSCILLATOR

Figure 5-15 shows the circuit of a tickler-type RF oscillator of the same type as that described previously, but with the battery replaced by two silicon solar cells PC_1 and PC_2. In bright sunlight, these two cells (International Rectifier S7M-C) deliver a total output of 6V and will adequately power the oscillator.

The output signal amplitude (open-circuit) is approximately 2V RMS. If a 365 pF variable capacitor is used for C_1, the coils described in Table 5-1 will permit tuning of the oscillator over the range extending from 440 kHz to 30 MHz in five bands

The MOSFET receives its gate-1 negative bias from the voltage drop resulting from the flow of drain current through source resistor R_4, and its gate-2 positive bias from voltage divider R_1–R_3. With an individual MOSFET, resistor R_1 in this divider may require some adjustment for maximum RF output voltage.

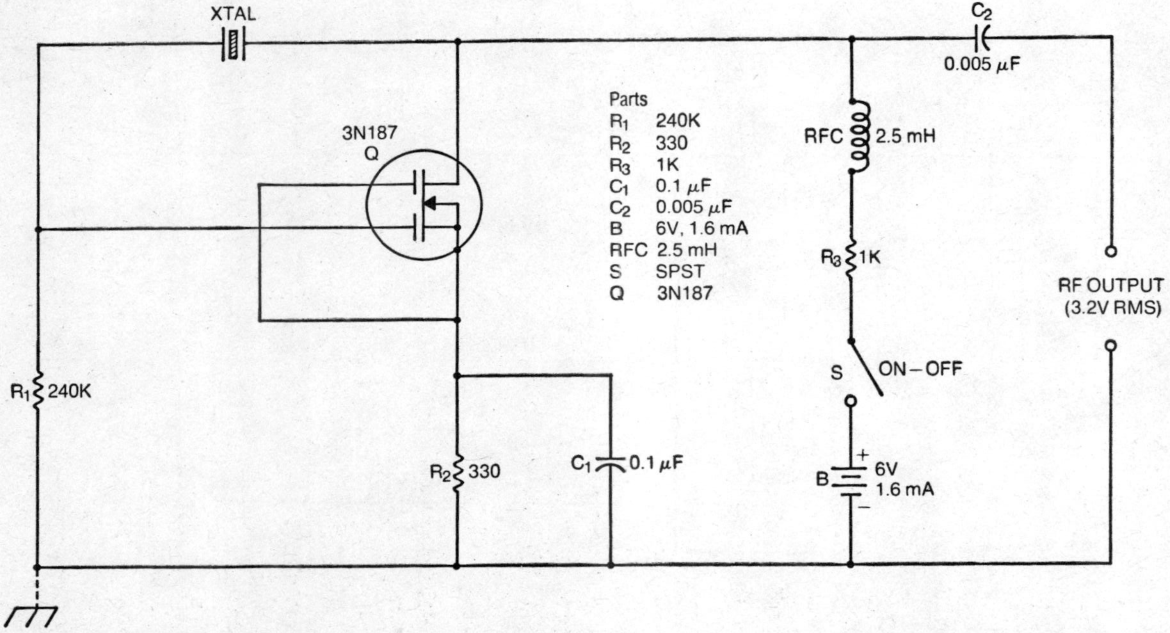

Fig. 5-14. Pierce crystal oscillator.

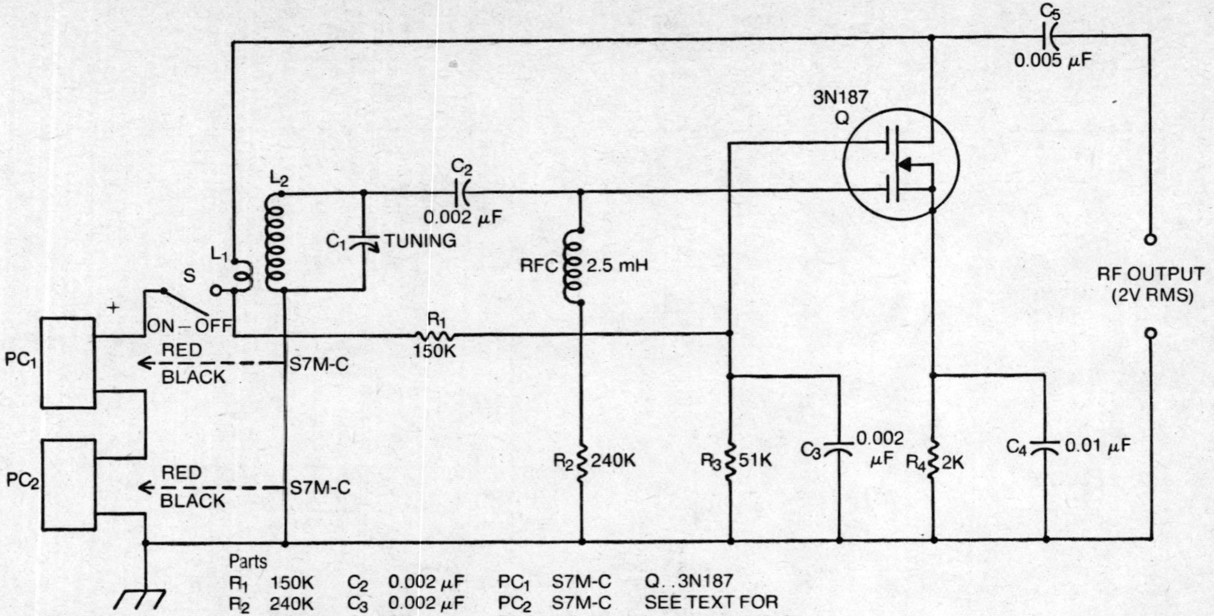

Parts

R_1	150K	C_2	0.002 μF	PC_1	S7M-C	Q. . 3N187
R_2	240K	C_3	0.002 μF	PC_2	S7M-C	SEE TEXT FOR
R_3	51K	C_4	0.01 μF	RFC	2.5 mH	L_1, L_2, AND C_1
R_4	2K	C_5	0.005 μF	S	SPST	

Fig. 5-15. Sun-powered RF oscillator.

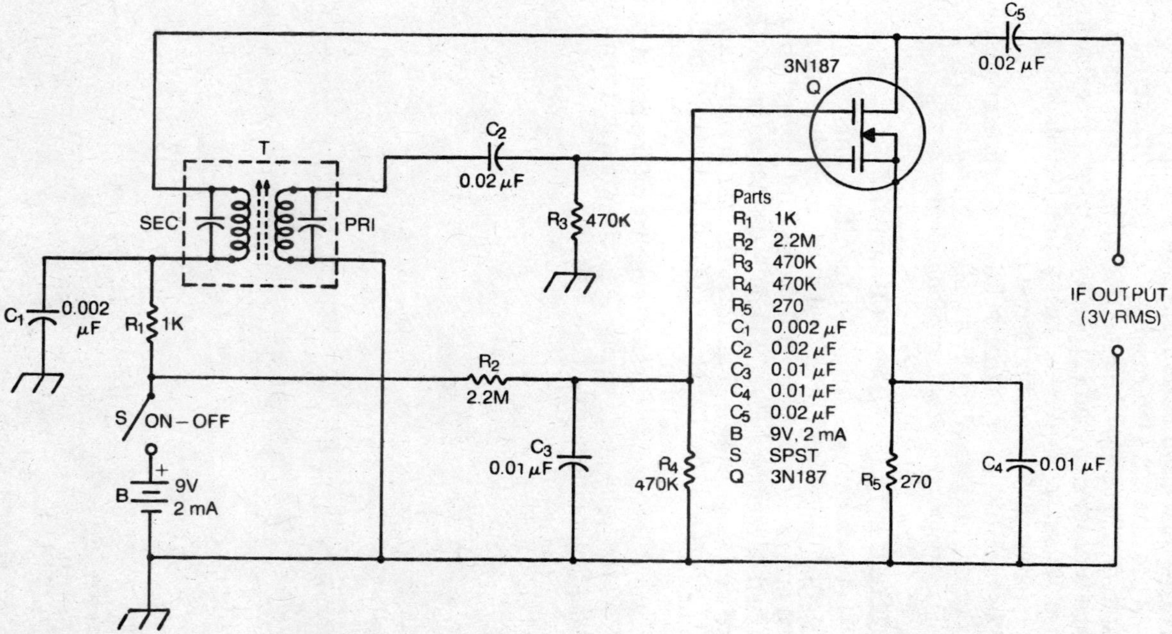

Fig. 5-16. Self-excited IF oscillator.

SELF-EXCITED IF OSCILLATOR

Figure 5-16 shows the circuit of a self-excited intermediate-frequency oscillator. This is a tickler-type oscillator with the two coils supplied by IF transformer *T*. The latter is chosen for the intermediate frequency of interest—for example, 455 kHz, 1500 kHz, or 10.7 MHz. The transformer must be connected in correct polarity for positive feedback; if the circuit does not oscillate when it is first tested, reverse the connections to *one* of the transformer coils.

The open-circuit IF output voltage is approximately 3V RMS. Current drain from the 9V supply is approximately 2 mA, but this may vary somewhat with individual MOSFETs.

CRYSTAL-CONTROLLED IF OSCILLATORS

If better stability is required than that afforded by the self-excited IF oscillator described in the preceding section, a crystal oscillator may be used. Either of the crystal circuits described earlier (see Fig. 5-13 and 5-14) can be used. If the Pierce circuit is chosen (Fig. 5-14), it is necessary only to insert a crystal oscillating at the desired intermediate frequency. If the standard circuit is used (Fig. 5-13), inductor *L* and variable capacitor C_3 can be supplied by one-half of an IF transformer, and the output can be taken from the coil–capacitor circuit of the other half.

Chapter 6

Test Instruments

The unique characteristics of the MOSFET make it particularly suitable for applications involving test instruments—expecially the miniaturized, battery-operated varitey. Important advantages obtained are negligible loading, reduced operating current, gow drift, high sensitivity, and small size.

This chapter presents 15 selected instrument circuits. Several circuits appearing elsewhere in this book and intended primarily for other applications may also be used in instrumentation. See, for example, Figs. 2-13 through 2-22, 3-1 through 3-3, 3-12, 4-6, 4-12 through 4-14, 4-16, 4-17, 5-1 through 5-8, 5-10 through 5-14, and 5-16.

In each of the circuits in this chapter, unless indicated otherwise on the diagram or in the text, capacitances are in picofarads, resistances in ohms, and inductances in henrys. Resistors are ½W, and electrolytic capacitors are rated at 25V. For simplicity, batteries are shown for DC supply; however, a *well filtered* power-line-operated power supply may be used instead.

Before undertaking the wiring or operation of any circuit in this chapter, read carefully the hints and precautions given in Chapter 1.

SINGLE-MOSFET ELECTRONIC DC VOLTMETER

Figure 6-1 shows a simple electronic DC voltmeter circuit offering an input resistance of 11M, adapted from a circuit by

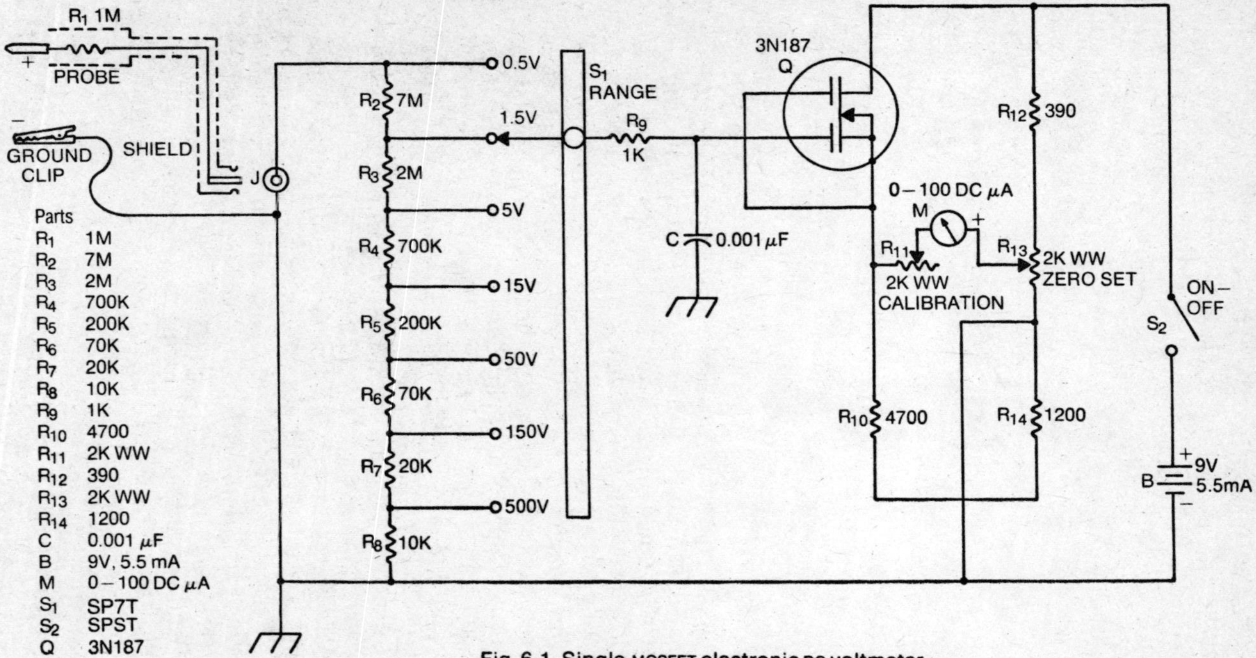

Parts

R_1	1M
R_2	7M
R_3	2M
R_4	700K
R_5	200K
R_6	70K
R_7	20K
R_8	10K
R_9	1K
R_{10}	4700
R_{11}	2K WW
R_{12}	390
R_{13}	2K WW
R_{14}	1200
C	0.001 μF
B	9V, 5.5 mA
M	0 – 100 DC μA
S_1	SP7T
S_2	SPST
Q	3N187

Fig. 6-1. Single-MOSFET electronic DC voltmeter.

Siliconix. This arrangement is seen to be similar to the vacuum-tube counterpart. Seven voltage ranges are employed between 0.5V full-scale and 500V full-scale.

As in most conventional electronic voltmeters—tube-type and transistorized—the indicating microammeter is connected in a bridge circuit of which the internal drain-to-source resistance of the MOSFET is a part. With zero-signal input to jack *J*, the meter is set to zero by balancing this bridge (potentiometer R_{13} is the balancing element). When a DC voltage subsequently is applied at jack *J*, the bridge unbalances (since the MOSFET drain resistance changes), and the meter reads proportionately.

The accuracy of the instrument depends in large part upon the precision of resistors R_1 to R_8 in the input circuit. Special close-tolerance instrument resistors must be obtained, or the odd values must be made up by connecting several lower values in series.

Initial calibration is straightforward:

(1) With zero voltage at the input probe, close switch S_2 and set potentiometer R_{13} to zero the pointer of meter *M* (range switch S_1 may be at any setting for this operation).

(2) Set range switch S_1 to its 1.5V position.

(3) Connect an accurately known 1.5V source to the input probe and ground clip.

(4) Adjust calibration control R_{11} for exact full-scale deflection of meter *M*.

(5) Temporarily remove the input voltage and note whether meter is still zeroed. If it is not, reset R_{13}.

(6) Work back and forth between steps 3, 4, and 5 until a 1.5V input deflects the meter to full scale, and the meter remains zeroed when the 1.5V input is removed. Potentiometer R_{11} will not need to be reset unless its setting is disturbed or the instrument is being periodically recalibrated. Zero-set potentiometer R_{13} will need to be reset before each period of use.

The *RC* filter formed by R_9 and *C* serves to remove any AC components and hash that enter the input section of the circuit. The 1M resistor R_1 in the shielded probe protects the circuit from hand capacitance.

Current drain from the 9V supply is approximately 5.5 mA, but this may vary somewhat with individual MOSFETs.

BALANCED ELECTRONIC DC VOLTMETER

The circuit shown in Fig. 6-2, like its tube or transistor counterpart, requires less zero resetting than a single-ended circuit does, if a pair of MOSFETs with matched characteristics

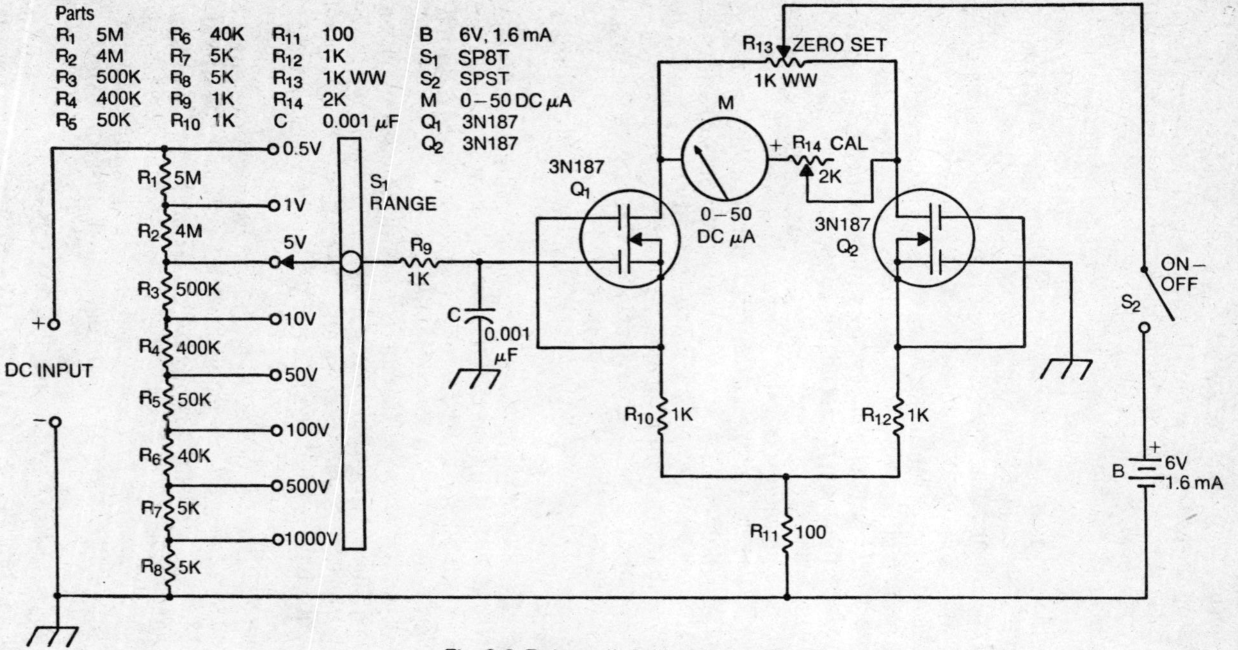

Fig. 6-2. Balanced electronic DC voltmeter.

is used. Any drift in MOSFET Q_1 is balanced out by comparable drift in Q_2; the internal drain circuits of the MOSFETs, in conjunction with resistors R_{10}, R_{11}, R_{12}, and R_{13}, form a bridge for this purpose and for the normal operation of the voltmeter. Indicating microammeter M is the detector in this bridge circuit. With zero signal at the input terminals, the meter is set to zero by balancing this bridge (potentiometer R_{13} is the balancing element). When a DC voltage subsequently is applied at the input terminals, the bridge unbalances (since the drain resistance changes), and the meter reads proportionately.

The input resistance of the instrument is 10M. Eight voltage ranges are provided between 0.5V full-scale and 1000V full-scale. The accuracy of the instrument depends in large part upon the precision of resistors R_1 to R_8 in the input circuit. Special close-tolerance instrument resistors having the exact values shown in Fig. 6-2 must be obtained, or the odd values must be made up by connecting several lower values in series.

Initial calibration is straightforward:

(1) With zero voltage at the input terminals, close switch S_2 and set potentiometer R_{13} to zero the pointer of meter M (range switch S_1 may be at any setting for this operation).

(2) Set range switch S_1 to its 1V position.

(3) Connect an accurately known DC source of 1V to the input terminals.

(4) Adjust calibration control R_{14} for exact full-scale deflection of meter M.

(5) Temporarily remove the input voltage and note whether or not the meter is still zeroed. If it is not, reset R_{13}.

(6) Work back and forth between steps 3, 4, and 5 until a 1V input deflects the meter to full-scale, and the meter remains zeroed when the 1V input is removed. Rheostat R_{14} will not need to be reset unless its setting is disturbed or the instrument is being periodically recalibrated. Zero set potentiometer R_{13} will not need frequent readjustment unless its setting is accidentally disturbed.

The *RC* filter formed by R_9 and C serves to remove any AC components and hash that enter the input section of the circuit. No further filtering or bypassing is required.

Current drain from the 6V supply is approximately 1.6 mA, but this may vary somewhat with individual MOSFETs.

CENTER-ZERO ELECTRONIC DC VOLTMETER

Figure 6-3 shows an electronic DC voltmeter circuit in which the indicating meter is initially set to place zero at the center of the scale. A positive input voltage then will deflect the meter up from center-scale, whereas a negative voltage

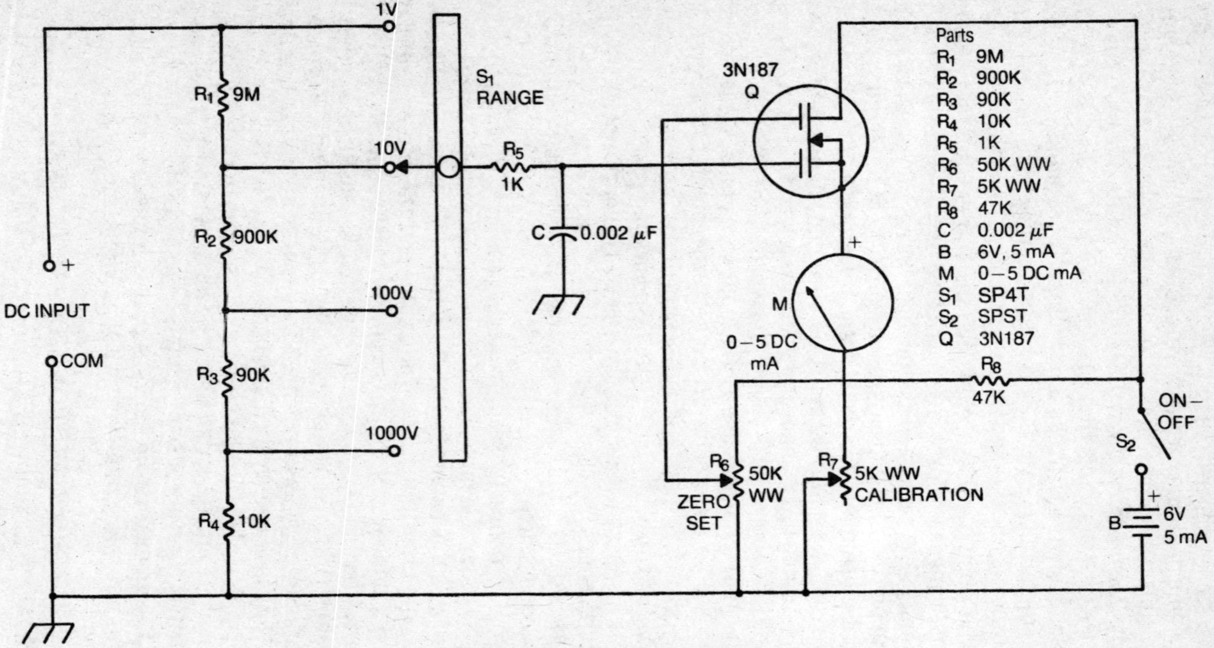

Fig. 6-3. Center-zero electronic DC voltmeter.

will deflect it downscale, and the input leads will not need to be swapped.

The input resistance of the instrument is 10M on all ranges. Four voltage ranges are provided between 1V full-scale and 1000V full-scale. The accuracy of the instrument depends largely upon the precision of resistors R_1 to R_4 in the input circuit. Special close-tolerance instrument resistors having the exact values given in Fig. 6-3 must be obtained, or the odd values must be made up by connecting several lower values in series.

The instrument is initially calibrated in the following manner:

(1) With zero voltage at the input terminals, close switch S_2 and set potentiometer R_6 to zero the pointer of meter M (range switch S_1 may be at any setting for this operation). Remember that the zero point is center-scale, i.e., at the 2.5 mA point on the scale of this 5 mA meter.

(2) Set range switch S_1 to its 1V position.

(3) Connect an accurately known source of +1V to the input terminals.

(4) Adjust calibration control R_7 for full upscale deflection of the meter.

(5) Reverse the DC source, applying −1V to the input terminals. The meter should now show full downscale deflection.

(6) Temporarily remove the input voltage and note whether or not the meter is still zeroed. If it is not, reset R_6.

(7) Work back and forth between steps 3, 4, 5, and 6 until a +1V input deflects the meter full-scale upward, a −1V input deflects it full-scale downward, and the meter remains zeroed when the DC input is removed. Rheostat R_7 will not need to be reset unless its setting is accidentally disturbed or the instrument is being periodically recalibrated. The reader must draw a special scale for the meter, having zero at the center.

The *RC* filter formed by R_5 and C serves to remove any AC components and hash that enter the input section of the circuit. No further filtering or bypassing is required.

Current drain from the 6V supply is approximately 5 mA, but this may vary somewhat with individual MOSFETs.

CRYSTAL-TYPE FREQUENCY STANDARD

Figure 6-4 shows the circuit of a 100 kHz frequency standard employing a Pierce crystal oscillator. The advantage of the Pierce circuit is its freedom from tuning; no adjustments are needed to start the oscillation—simply plug in the crystal and close the power switch.

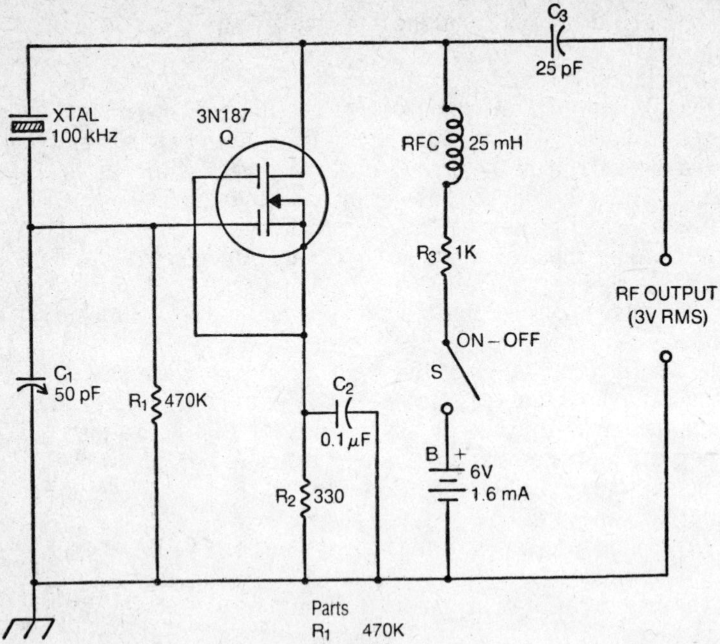

Parts	
R_1	470K
R_2	330
R_3	1K
C_1	50 pF
C_2	0.1 μF
C_3	25 pF
B	6V, 1.6 mA
RFC	25 mH
S	SPST
Q	3N187

Fig. 6-4. Crystal-type frequency standard.

In this arrangement, C_1 is a small trimmer-type capacitor which allows the frequency to be varied a few cycles on either side of 100 kHz for standardizing against WWV transmissions or some other accurate source.

The signal amplitude at the output terminals is approximately 3V RMS. Current drain from the 6V supply *B* is approximately 1.6 mA, but this may vary somewhat with individual MOSFETS.

SELF-EXCITED FREQUENCY STANDARD

A self-excited 100 kHz oscillator is satisfactory for use as a secondary standard and spotting oscillator when the very high stability and accuracy of a crystal-controlled oscillator is not required. Figure 6-5 shows a self-excited circuit of the tickler type.

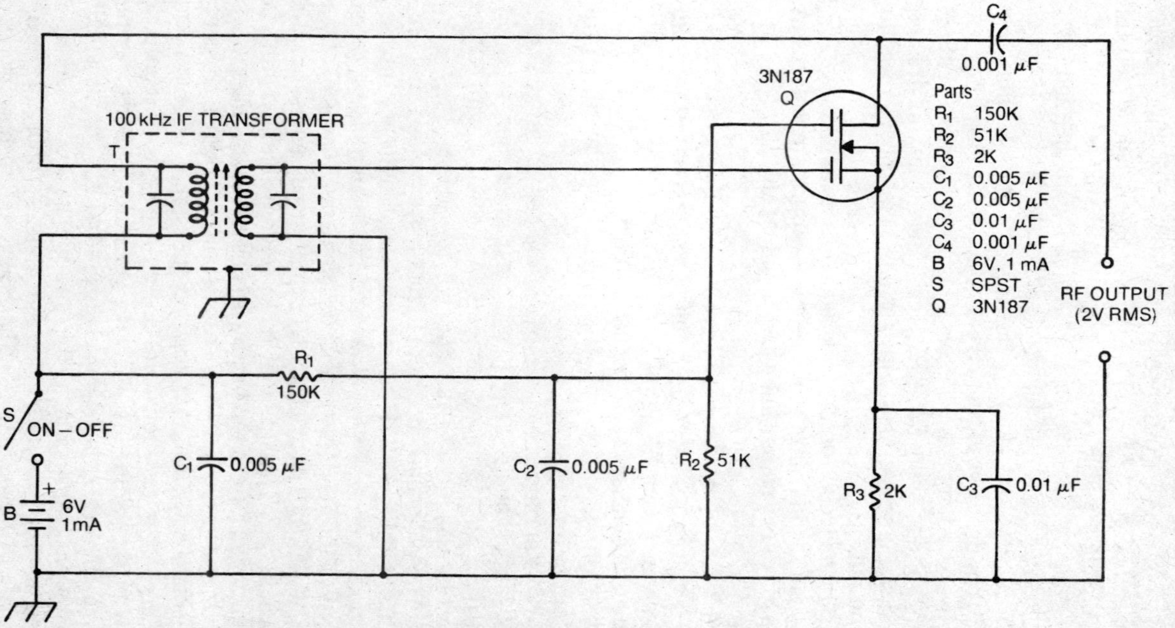

Fig. 6-5. Self-excited frequency standard.

In this arrangement, the signal and tickler windings both are supplied by 100 kHz intermediate-frequency transformer *T* (Miller No. 1890-P4). This is a slug-tuned transformer, and the slugs may be adjusted to zero-beat the oscillator with WWV transmissions. At this low frequency, the circuit is surprisingly stable. Polarity of the transformer is important; if oscillation does not occur, reverse the connections of either the primary *or* the secondary (not both).

The MOSFET receives its negative gate-1 bias from the voltage drop resulting from the drain current through RF-bypassed source resistor R_3, and its positive gate-2 bias from voltage divider R_1-R_2. Resistance R_1 in this divider may need some adjustment with an individual MOSFET for highest RF output.

Open-circuit output voltage is approximately 2V RMS. Current drain from the 6V supply is approximately 1 mA, but this may vary somewhat with individual MOSFETs.

AF–RF SIGNAL TRACER

Figure 6-6 shows the circuit of a signal tracer for following either an AF or RF signal through an electronic system, and which provides either visual indication (meter) or aural indication (loudspeaker). High input impedance is provided for either mode.

This arrangement consists of the combination MOSFET–IC ¼W-output audio amplifier described earlier (see *MOSFET Input for IC Amplifier*, Chapter 2), provided with AF and RF probes and a metering circuit (D_2-R_2-M).

The AF probe consists of shielded envelope and shielded cable with a 0.1 μF series capacitor. The RF probe is a demodulator type for following a modulated signal through a circuit, and it consists of shunt demodulator diode D_1, input capacitor C_1, load resistor R_1, and output envelope-coupling capacitor C_2. This probe also is shielded. Shielded input jack *J* is provided for connecting the probes to the amplifier.

When switch *S* is in position *1*, the speaker is connected to the output of the amplifier and the signal is made audible. When *S* is in position 2, a rectifier-type meter (composed of 0–1 DC milliammeter *M*, 1N34A germanium diode D_2, and 10K wirewound rheostat R_2) is connected to the output winding of the output transformer in the amplifier. The diode rectifies the output signal and deflects the meter proportionately, thus providing a visual signal. Rheostat R_2 is set to keep the meter from exceeding full-scale deflection on strong test signals.

The amplifier and, accordingly, the signal tracer itself, requires a 6V DC supply. The zero-signal current drain is 14

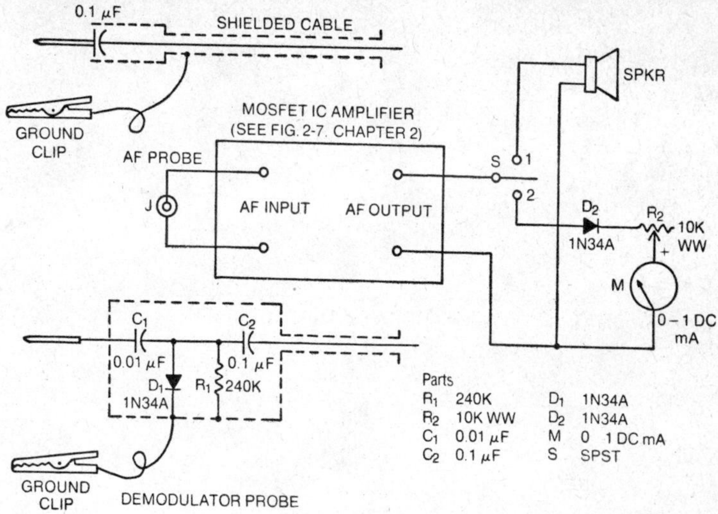

Fig. 6-6. Signal tracer (AF–RF).

mA, and the maximum-signal current drain is 87 mA; but these may vary somewhat with individual MOSFETs and ICs.

TUNED RF–IF SIGNAL TRACER

While aperiodic signal tracers such as the one described in the preceding section are easy to build and simple to use, professional signal tracing—especially in the RF and IF stages of radio and television receivers—often requires that the instrument be closely tuned to the traced signal. Tuned signal tracers can be very complicated. Figure 6-7, however, shows a fairly simple circuit that gives good selectivity and sensitivity. An electronic DC voltmeter (either vacuum-tube or transistorized) is connected to its output terminals to serve as the visual indicator.

The input of the circuit is of high impedance and untuned, the signal being coupled into the circuit through capacitor C_1 and resistor R_1. The signal is picked up with a shielded probe (of the simple test-meter type) plugged into jack J. The output of the circuit is tuned by means of C_5 and L. A set of five plug-in coils may be wound for L to cover a tuning range from 440 kHz to 30 MHz with the 365 pF variable capacitor C_5. Table 5-1 in Chapter 5 gives winding instructions for these coils (in using this table, take the L_2 data for each coil; ignore the L_1 values, since the signal tracer coil will not have a tickler).

Output of the tuned circuit is coupled through capacitor C_6 to a shunt rectifier circuit (germanium diode D and load

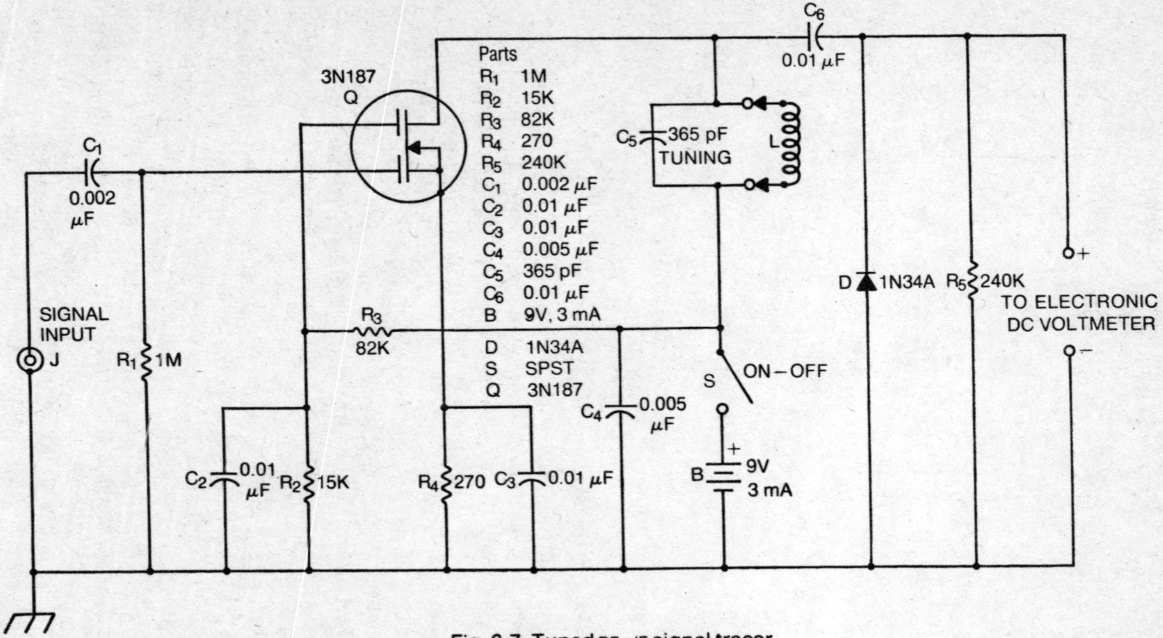

Fig. 6-7. Tuned RF—IF signal tracer.

resistor R_5). The resulting DC output of the rectifier, at the output terminals, drives the external voltmeter. When using the tracer, variable capacitor C_5 is simply tuned for maximum deflection of the meter.

In this circuit, the MOSFET receives its negative gate-1 bias from the voltage drop resulting from the flow of drain current through source resistor R_4, and its positive gate-2 bias from voltage divider R_2–R_3. In this divider, resistance R_3 may need some adjustment for maximum output. Current drain from the 9V supply is approximately 3 mA, but this may vary somewhat with individual MOSFETs.

AF SIGNAL-TRACER ADAPTER FOR AC VOLTMETER

The 2-MOSFET circuit shown in Fig. 6-8 allows a low-resistance AC voltmeter (1000Ω or 5000Ω per volt) to be used for audio-frequency signal tracing. The input impedance seen by the signal source is 0.1 μF (C_1) in series with 1M (R_1). The first stage is a common-source *RC* amplifier; the second stage is a source follower delivering output to the meter from 500Ω resistor R_7. Overall voltage gain of the system is approximately 15 (that is, 30 for the input stage and 0.5 for the output stage). With sensitivity control R_1 set for maximum sensitivity, the maximum input signal at jack *J* before peak clipping in the output is approximately 50 mV RMS. The corresponding signal at the output terminals is approximately 0.75V RMS. This provides a good deflection on the 0–1.5V range of the AC voltmeter connected to these terminals. A simple shielded probe may be plugged into jack *J*.

In this circuit, MOSFET Q_1 receives its negative gate-1 bias from the voltage drop resulting from the flow of drain current through source resistor R_4, and its positive gate-2 bias from voltage divider R_2–R_3. In this divider, resistance R_3 may require some adjustment for maximum signal output. The other MOSFET, Q_2, receives its negative gate-1 and gate-2 bias from the voltage drop resulting from the flow of drain current through source resistor R_7. Current drain from the 12V supply is approximately 4 mA, but this may vary somewhat with individual MOSFETs.

ACTIVE PROBE

Figure 6-9 shows the circuit of a probe (useful with such instruments as oscilloscopes and meters) that provides not only voltage gain (10, open-circuit), but also high input impedance (1 μF capacitor C_1 in series with 10M resistor R_1) and wideband frequency response (−3 dB at 50 Hz and −6 dB at 10 MHz). This circuit must be carefully shielded and its leads individually dressed, all according to wideband practice.

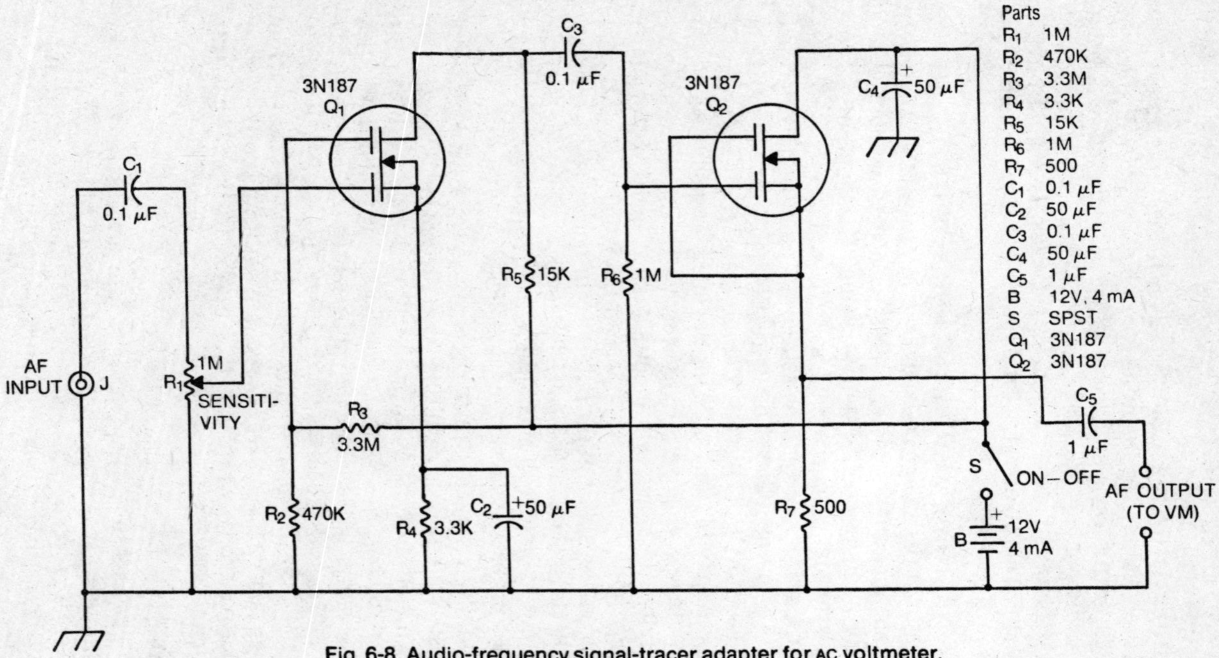

Fig. 6-8. Audio-frequency signal-tracer adapter for AC voltmeter.

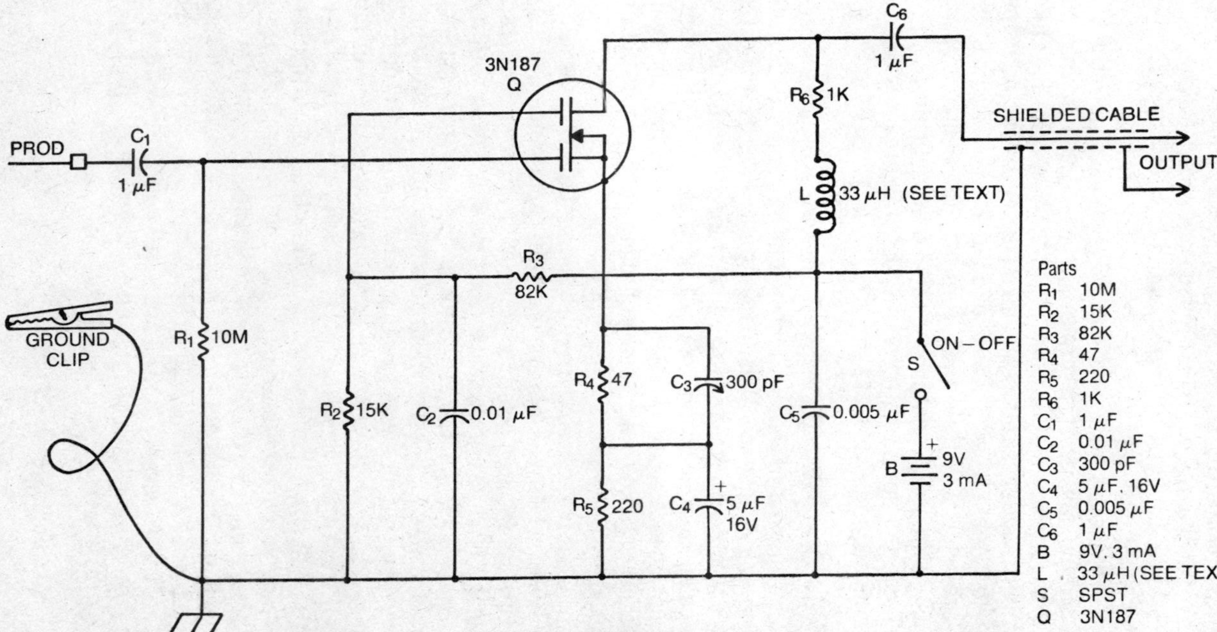

Parts	
R_1	10M
R_2	15K
R_3	82K
R_4	47
R_5	220
R_6	1K
C_1	1 μF
C_2	0.01 μF
C_3	300 pF
C_4	5 μF, 16V
C_5	0.005 μF
C_6	1 μF
B	9V, 3 mA
L	33 μH (SEE TEXT)
S	SPST
Q	3N187

Fig. 6-9. Active probe.

The 33 μH inductor L is a Miller 9250-333 molded RF choke. It provides compensation in this video amplifier circuit, as does the 300 pF miniature trimmer capacitor C_3. The latter must be rigidly mounted in such a position that its adjusting screw is easily reached through a hole in the probe housing. Set C_3 for best overall frequency response.

In this circuit, the MOSFET receives its negative gate-1 bias from the voltage drop resulting from the flow of drain current through the source-resistor combination R_4–R_5 (bypassed separately for both low and high frequencies), and its positive gate-2 bias from voltage divider R_2–R_3. In this divider, resistance R_3 may need some adjustment for minimum distortion in the output signal. Current drain from the 9V supply is approximately 3 mA, but this may vary somewhat with individual MOSFETs.

WIDE-RANGE LC CHECKER

Figure 6-10 shows the circuit of a checker which allows a variable-frequency audio generator (tuning from 20 Hz to 20 kHz) to be used to measure inductance from 6.3 mH to 6329H and capacitance from 0.0632 μF to 63,291 μF Other ranges for capacitance may be obtained by changing the value of inductor L, and other ranges of inductance may be obtained by changing the value of capacitor C_2.

The MOSFET provides a source-follower circuit which drives a series-resonant circuit containing the unknown capacitance connected in series with a known inductance, L, or the unknown inductance connected in series with a known capacitance, C_2. The generator is tuned for peak deflection of meter M, which is part of an AC milliammeter connected in the series-resonant circuit. At that point, the generator frequency is noted, and the unknown component calculated in terms of the known values. Thus, $C_x = 1/39.5f^2L_s$ and $L_x = 1/39.5f^2C_s$. The unknown component is connected to terminals X–X. Resonant current flowing through 100Ω resistor R_3 develops a voltage drop across this resistor, and this voltage is rectified by diode D to deflect meter M. Standard inductor L is a 1 mH slug-tuned coil (Miller 22A103RBI) which may be set exactly to 1 mH in a suitable bridge. Standard capacitor C_2 is a 0.01 μF silver-mica unit which must be obtained with as high accuracy as practicable.

Capacitance Measurement

To use the instrument for determining capacitance values, follow this procedure:

(1) Connect the signal generator to the AF input terminals.

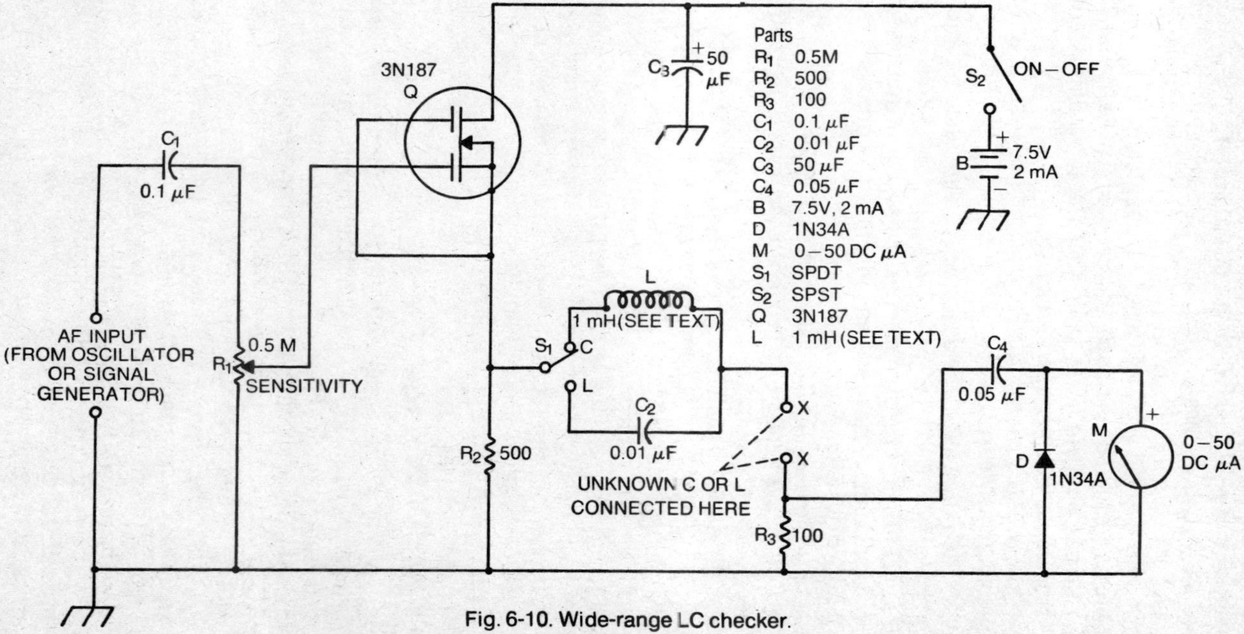

Fig. 6-10. Wide-range LC checker.

(2) Connect the unknown capacitor to terminals X–X.

(3) Set switch S_1 to C.

(4) Close switch S_2 and turn on generator.

(5) Tune generator throughout its range, watching for a peak deflection of meter M.

(6) At peak, read frequency f from generator dial.

(7) Calculate the capacitance: $C = 1/0.0395f^2$, where C is in farads and f in hertz. Note that the formula is simplified, since the inductance of L is constant.

Inductance Measurement

The following steps describe the sequence for measuring inductance.

(1) Connect the signal generator to the AF input terminals.

(2) Connect the unknown inductor to terminals X–X.

(3) Set switch S_1 to L.

(4) Close switch S_2 and turn on generator.

(5) Tune generator throughout its range, watching for a peak deflection of meter M.

(6) At peak, read frequency f from generator dial.

(7) Calculate the inductance: $L = 1/3.95f^2 \times 10^{-7}$, where L is in henrys and f in hertz. Note that the formula is simplified, since the capacitance of C_2 is constant.

In this circuit, the MOSFET receives its negative gate-1 and gate-2 bias from the voltage drop resulting from the flow of drain current through source resistor R_2. Current drain from the 7.5V supply is approximately 2 mA, but this may vary somewhat with individual MOSFETs.

AC NULL DETECTOR

A sharply tuned null detector is an asset in balancing an AC bridge, and it is important also that the input impedance of the detector be high. Figure 6-11 shows a null-detector circuit meeting these specifications. This arrangement consists of the RC-tuned peak amplifier described in Chapter 2, followed by a simple AC meter circuit (C–D_1–D_2–R–M).

For the common 1000 Hz bridge frequency, the frequency-determining resistances and capacitances in the amplifier (see Fig. 2-18) are: $C_2 = C_3 = 0.1\ \mu F$, $C_4 = 0.2\ \mu F$, $R_3 = R_5 = 1592\Omega$, and $R_4 = 796\Omega$.

The gain control in the amplifier can be used to set the prenull deflection of meter M to full-scale. Rheostat R in the metering circuit is preset for full-scale deflection of meter M when the amplifier gain control is in its maximum-sensitivity position.

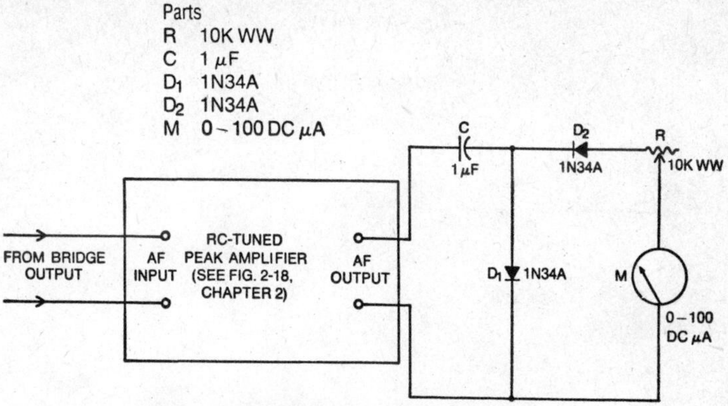

Fig. 6-11. Alternating-current null detector.

SENSITIVE LIGHT METER

Figure 6-12 shows the circuit, which increases the sensitivity of a self-generating selenium photocell. The circuit is basically that of a MOSFET DC amplifier which drives a 0–100 DC microammeter. The meter is connected in a 4-arm bridge circuit in the output of the MOSFET. The arms of this bridge consist of resistors R_2, R_3, and R_4 and the internal drain-to-source resistance of the MOSFET.

Initially, with photocell *PC* (International Rectifier B3M-C1) darkened, the bridge is balanced and the meter, therefore, zeroed by adjustment of rheostat R_4. For this operation, potentiometer R_1 should be set to its maximum-sensitivity position. When subsequently the photocell is illuminated, the resulting output voltage drives the MOSFET gates positive; and this changes the internal resistance of the MOSFET, unbalancing the bridge and causing the meter to deflect proportionately.

Illumination of 10 footcandles results in full-scale deflection of the meter. By comparison, if the photocell is connected directly to the microammeter, illumination of 100 footcandles will give approximately 75% of full-scale deflection. If facilities for calibration are available, a dial attached to potentiometer R_1 may be graduated to read directly in footcandles.

Current drain from the 6V supply is approximately 3 mA, but this may vary somewhat with individual MOSFETs.

SINE- TO SQUARE-WAVE CONVERTER

Figure 6-13 shows an aperiodic (untuned) circuit for converting input sine waves into output square waves. This

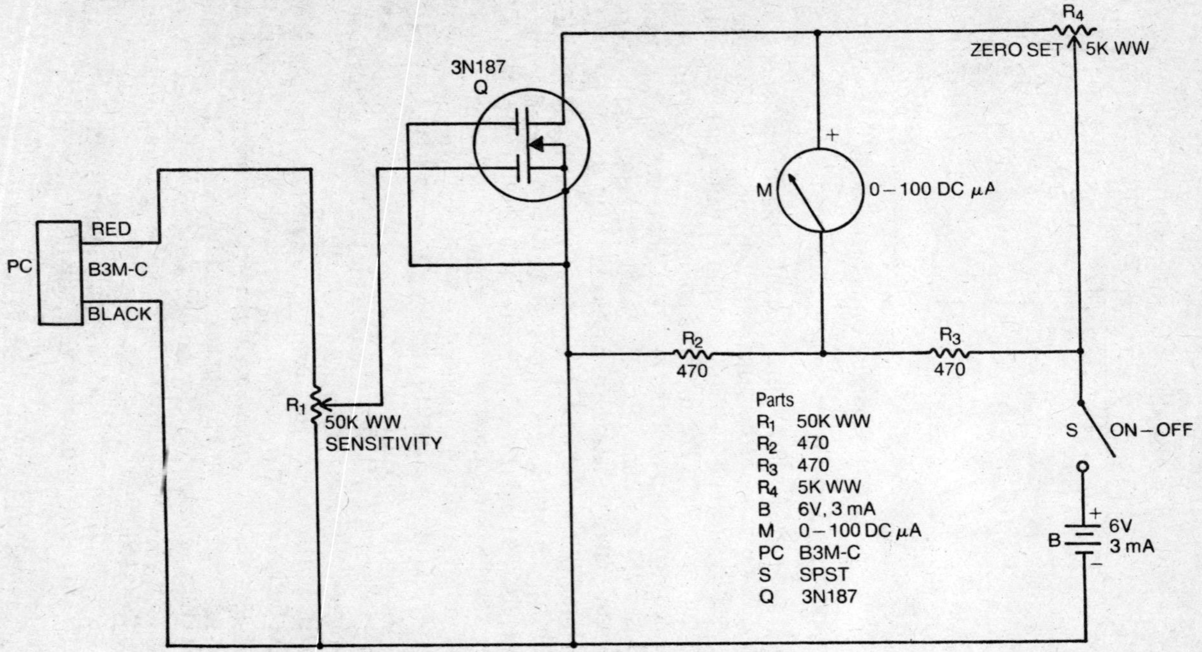

Parts	
R_1	50K WW
R_2	470
R_3	470
R_4	5K WW
B	6V, 3 mA
M	0 – 100 DC μA
PC	B3M-C
S	SPST
Q	3N187

Fig. 6-12. Sensitive light meter.

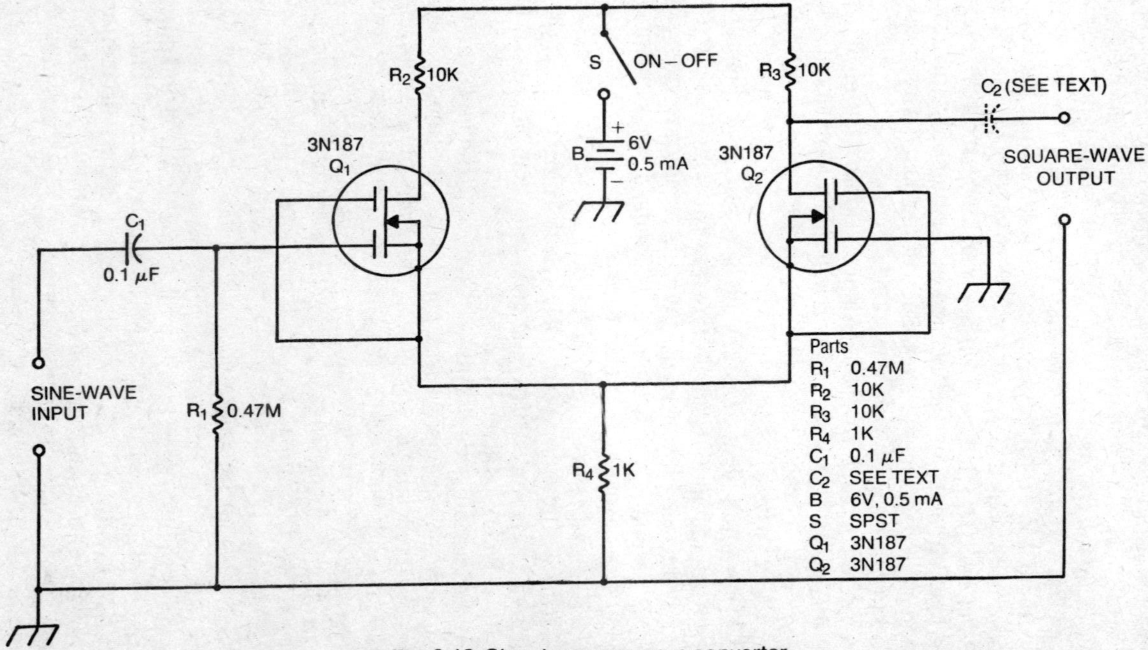

Fig. 6-13. Sine- to square-wave converter.

arrangement consists essentially of cascaded overdriven stages coupled by common source resistor R_4.

The minimum input signal amplitude for good square-wave shape is 0.8V RMS. In some instances, especially where the sine-wave signal source has low output, the input voltage of the converter may be derived from an intermediate amplifier stage (see, for example, Fig. 2-1, Chapter 2).

Capacitor C_2 will be required when the output load device would short-circuit the drain of MOSFET Q_2. The actual capacitance will depend upon the resistance and reactance of the load; it must be determined experimentally, that value being chosen which will least distort the square waveform. In most instances, 1 μF will be suitable.

Current drain from the 6V supply is approximately 0.5 mA, but this may vary somewhat with individual MOSFETs.

AM MONITOR

Figure 6-14 shows the circuit of a sensitive monitor for amplitude-modulated signals. Operable at some distance from the transmitter, the device receives its signal through a small pickup antenna, which may be a standard whip antenna or simply a vertical length of stiff wire or rod.

The signal is tuned in with inductor L and capacitor C_2. For general-coverage tuning from 440 kHz to 30 MHz, C_2 is a 365 pF variable capacitor, and plug-in coils may be wound according to the instructions given in Table 4-1, Chapter 4.

The signal is demodulated by 1N34A diode D, and the resulting AF envelope, corresponding to the modulation, is presented to gate 1 of the MOSFET through volume control R_2. The MOSFET functions here solely as an AF amplifier, providing the sensitivity of the monitor.

A high-impedance headset or earpiece may be connected directly to the output terminals; or, for a louder signal, a suitable amplifier and speaker (see, for example, Fig. 2-7, Chapter 2) may be connected.

In this circuit, the MOSFET receives its negative gate-1 bias from the voltage drop resulting from the flow of drain current through source resistor R_5, and its positive gate-2 bias from voltage divider R_3–R_4. Resistance R_4 in this divider may require some adjustment for maximum output. Current drain from the 9V source is approximately 2 mA, but this may vary somewhat with individual MOSFETs.

CW MONITOR

Figure 6-15 shows the circuit of a sensitive monitor for CW signals. Operable some distance from the transmitter, the

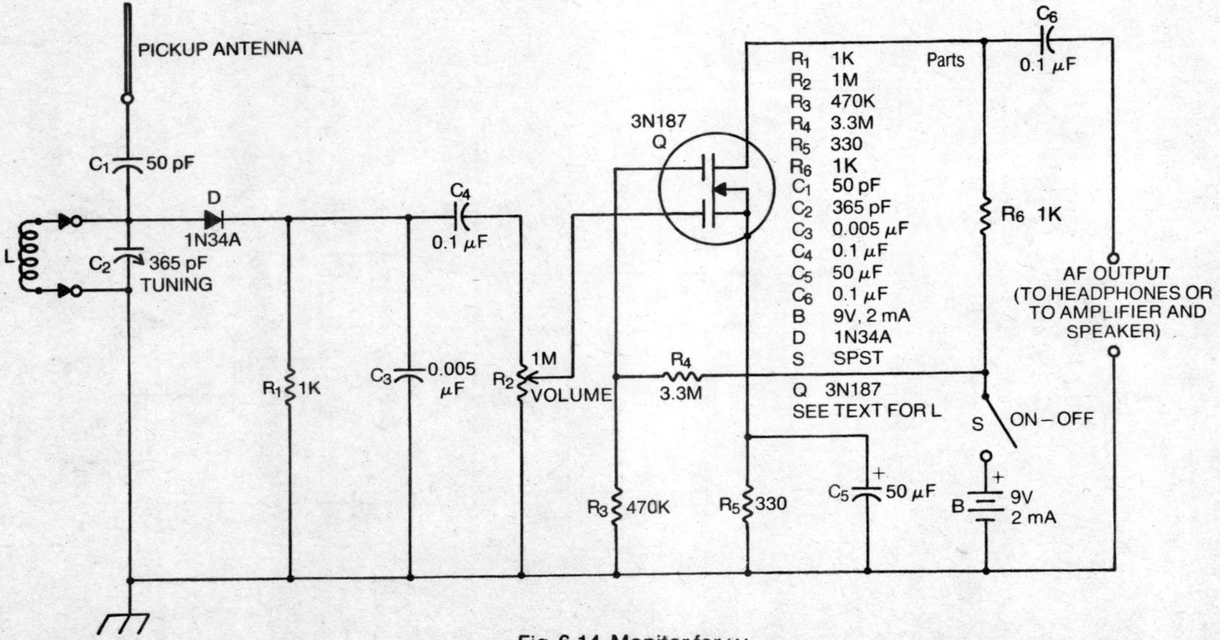

Fig. 6-14. Monitor for AM.

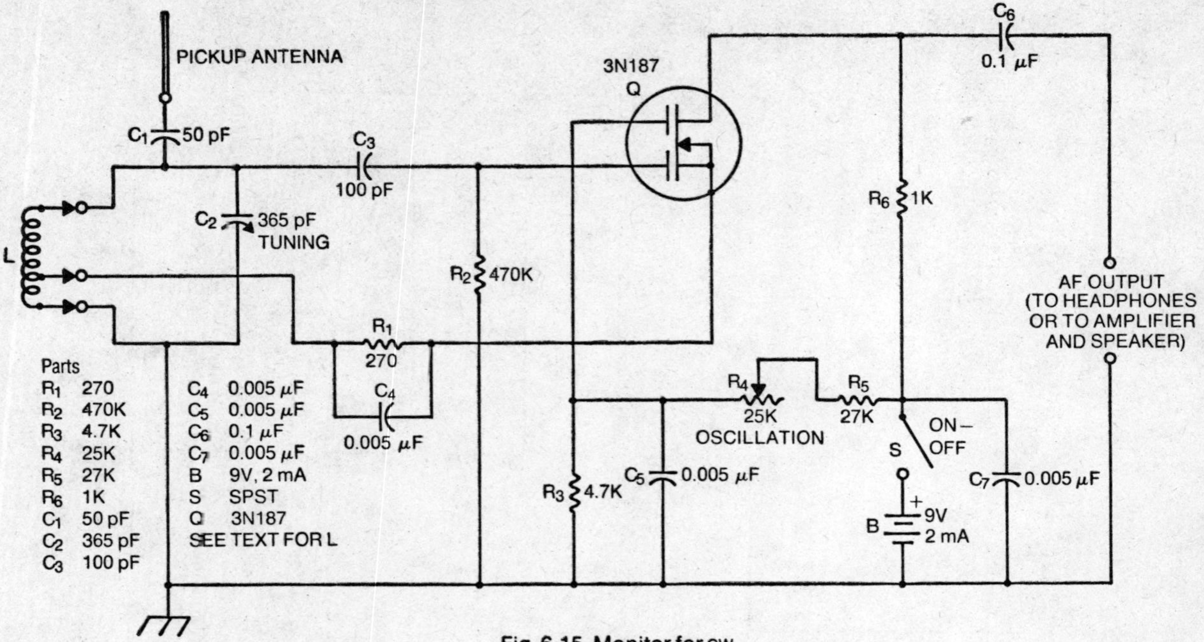

Fig. 6-15. Monitor for cw.

device is essentially an oscillating detector which receives its signal through a small pickup antenna. The latter may be a standard whip antenna or a vertical rod or wire.

The signal is tuned in with inductor L and capacitor C_2. For general-coverage tuning from 440 kHz to 30 MHz, C_2 is a 365 pF variable capacitor, and plug-in coils may be wound according to the instructions given in Table 5-2, Chapter 5.

In this circuit, oscillation (regeneration) is controlled by adjustment of rheostat R_4 in the voltage divider which supplies positive bias to gate 2 of the MOSFET. This control is set for the most pleasing tone. Gate 1 receives its negative bias from the voltage drop resulting from the flow of drain current through the RF-bypassed source resistor R_1. High-resistance headphones may be connected directly to the output terminals; for louder signals, an amplifier and speaker (see, for example, Fig. 2-7, Chapter 2) may be connected instead.

Current drain from the 9V supply is approximately 2 mA, but this may vary somewhat with individual MOSFETs.

Chapter 7

Miscellaneous Circuits

The 15 additional circuits presented in this chapter do not fit precisely into the categories covered by the 6 preceding chapters. Each circuit exploits one or more of the desirable properties of the MOSFET: high input impedance, high transconductance, negligible cross modulation, low feedback capacitance. These circuits will suggest other applications to the alert experimenter.

In each of the circuits, unless indicated otherwise on the diagram or in the text, capacitances are in picofarads, resistances in ohms, and inductances in henrys. Resistors are ½W, and electrolytic capacitors are to be rated at 25V. For simplicity, batteries are shown for DC supply; however, a *well filtered* power supply may be used instead of a battery. It is assumed that all radio-frequency circuits will be adequately shielded and will be wired with the shortest practicable direct leads.

Before undertaking the wiring or testing of any circuit in this chapter, read carefully the hints and precautions given in Chapter 1.

CONSTANT-CURRENT ADAPTER

Figure 7-1A shows the circuit of a current regulator based upon the almost flat drain-current characteristic of the MOSFET. A constant-current device of this type tends to hold the current steady while the applied voltage or load resistance varies. As shown in Fig. 7-1B, the current through the load

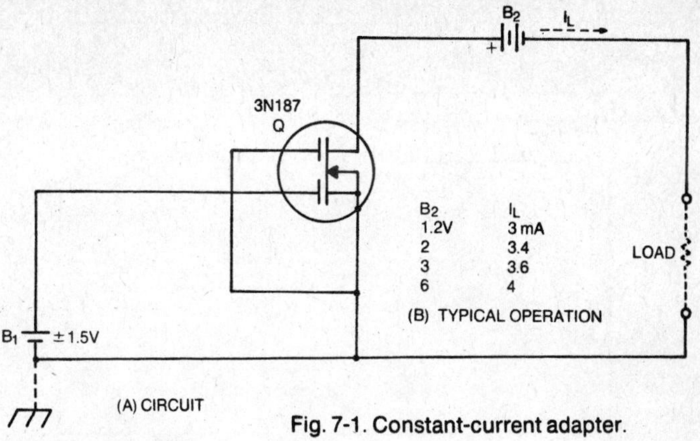

Fig. 7-1. Constant-current adapter.

(here a low resistance) varies only 1.33 to 1 as the applied voltage varies 5 to 1.

In this arrangement, the MOSFET (gate 1) is prebiased by the 1.5V cell B_1. The internal drain-to-source circuit then acts as an automatic resistor whose value changes in the proper direction and amount to slow current changes in the drain circuit containing battery B_2 and the load. When B_2 is 1.2V, the load current I_L is 3 mA; and when B_2 is 6V, I_L is 4 mA (see Fig. 7-1B). The B_1 voltage is selected in any special application of this adapter to place the operating point along the flat part of the drain-current–drain-voltage characteristic between the limits of the low and high values of B_2.

Since the current drain from B_1 is infinitesimal, no switch is generally needed for this cell.

CHARGE DETECTOR (ELECTROSCOPE)

A device for detecting the presence of static charges is indispensable in many areas of electronics. The old-fashioned gold-leaf electroscope has a long record of usefulness as a static detector, but this device must be held upright at all times. The device shown schematically in Fig. 7-2, however may be held in any position. No contact need be made with the charged body; the disc is merely held near it.

This arrangement is similar to an electronic DC voltmeter, but with an open input circuit. Gate 2 of the MOSFET is floating and thus is highly susceptible to nearby electric charges. The gate is connected to a smooth, ½ in. diameter metal disc on the end of a rod or short length of stiff wire (a polished metal ball will also serve), and this disc is held in the suspected field or close to the charged body. The picked-up energy is amplified

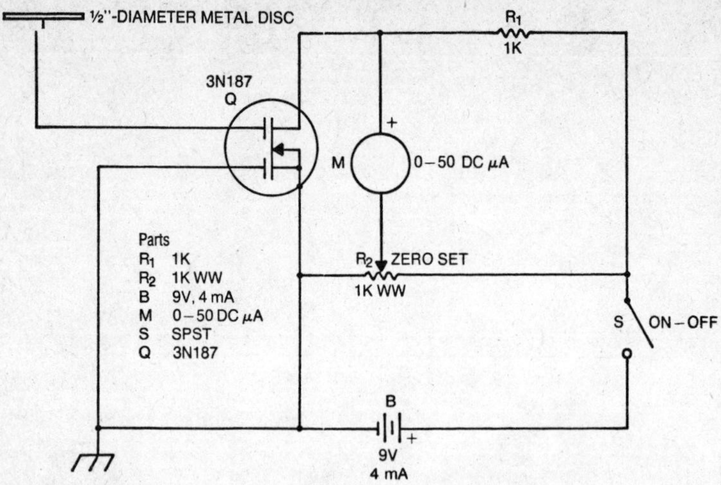

Fig. 7-2. Charge detector (electroscope).

by the MOSFET, and the latter drives miniature DC microammeter M.

As in an electronic voltmeter circuit, the indicating meter is connected in a 4-arm bridge circuit in the output section. The arms of this bridge consist of resistor R_1, the two "halves" of potentiometer R_2, and the internal drain-to-source resistance of the MOSFET. With the pickup disc clear of any electric fields and pointed away from the operator's body, potentiometer R_2 is set to zero the meter. The device will then respond by an upward deflection of the meter whenever the disc is moved into a field. A substantial deflection is obtained, for example, when the pickup is pointed at the high-voltage wiring in a TV receiver.

Current drain from the 9V battery is approximately 4 mA, but this may vary somewhat with individual MOSFETs.

FLIP-FLOP

Figure 7-3 shows the circuit of a conventional flip-flop adapted to MOSFETs. The circuit switches with clean precision at rates up to 2 kHz and delivers a reasonably square waveform. The open-circuit output swings to approximately 9V peak in response to a 9V negative pulse at the input terminals. In this arrangement, steering diodes D_1 and D_2 and commutating capacitors C_2 and C_3 should not be omitted. In setups in which the circuit into which the flip-flop feeds will ground the drain of MOSFET Q_2, a blocking capacitor C_4 must be included. The capacitance of this component will most often be

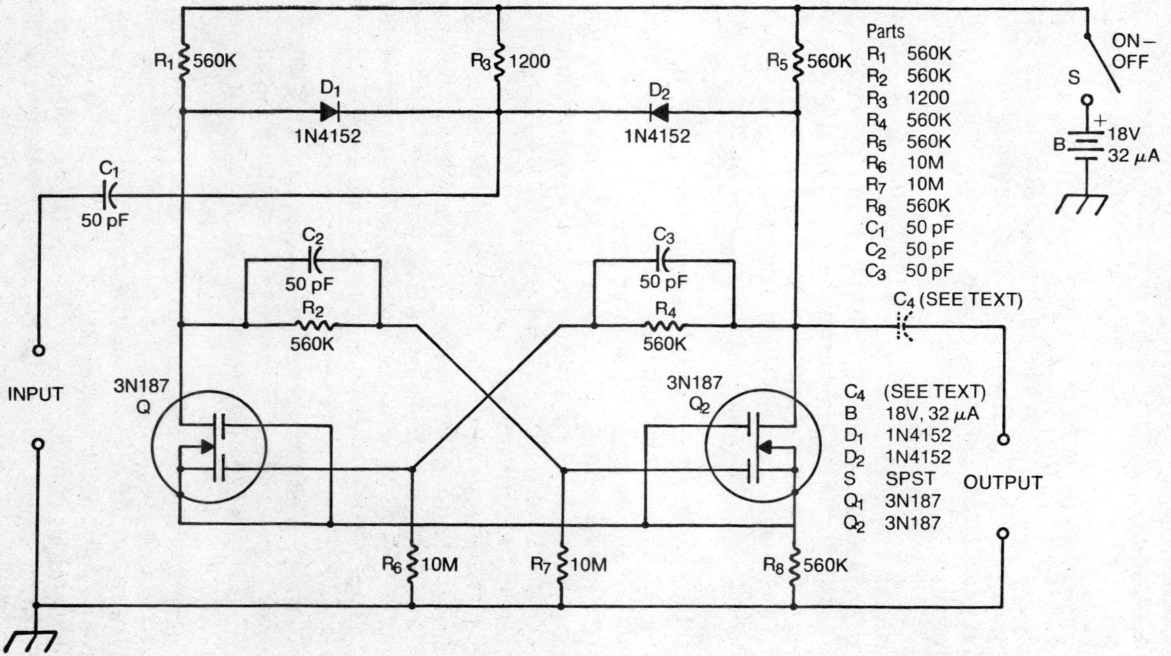

Fig. 7-3. Flip-flop.

100 pF, but must be chosen for least distortion of the flip-flop output and maximum transfer of energy to the driven circuit.

Some advantage is to be gained with a matched pair ot MOSFETs. Current drain from the 18V supply is approximately 32 μA, but this may vary somewhat with individual MOSFETs.

SUPERSONIC PICKUP

Figure 7-4 shows a pickup—preamplifier circuit which may be built into a small head for positioning close to a supersonic sound source. Since the current drain is low, a self-contained battery may be expected to give long service. The open-circuit voltage gain is approximately 30. The maximum transducer (microphone) signal, with gain control potentiometer R_1 set for maximum gain, is approximately 50 mV RMS before peak clipping in the amplifier output signal. The corresponding maximum output signal is approximately 1.5V RMS. In the construction of this device, rigid mounting should be employed to prevent vibration at supersonic frequencies.

The MOSFET receives its negative gate-1 bias from the voltage drop resulting from the flow of drain current through source resistor R_4, and its positive gate-2 bias from voltage divider R_2–R_3. With an individual MOSFET some adjustment of resistance R_3 may be required for maximum signal output.

BALANCED MODULATOR

A balanced modulator is required in sideband communications and in heterodyne-type instruments utilizing carrier-elimination techniques. The triode-type balanced modulator is superior in some respects to the simplc diode modulator. Figure 7-5 shows the circuit of a 2-MOSFET balanced modulator of the triode type.

In this circuit, the carrier component is applied to the RF input terminals and arrives at the gate-1 electrodes in parallel, whereas the modulation component (applied at the AF input terminals) arrives at the gate-1 electrodes in push-pull. Ideal symmetry of the circuit is established by the close matching of MOSFETs and circuit components in each of its halves. Close adjustment of the dual variable capacitor C_3–C_4 and of potentiometer R_7 then results in elimination of the RF component in transformer T_3. When the AF component is applied, it modulates the carrier; but the latter has been suppressed, so only the two resulting sidebands are transmitted by T_3. In both communications and instrumentation, one of these sidebands is then amplified by a selective filter following the balanced modulator.

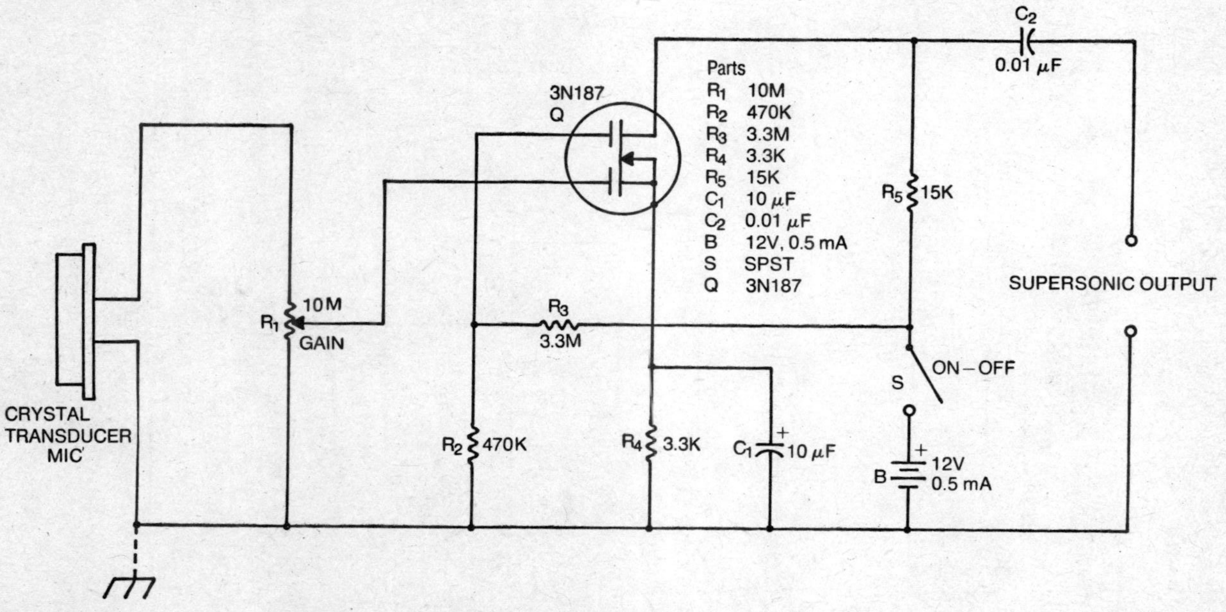

Fig. 7-4. Supersonic pickup.

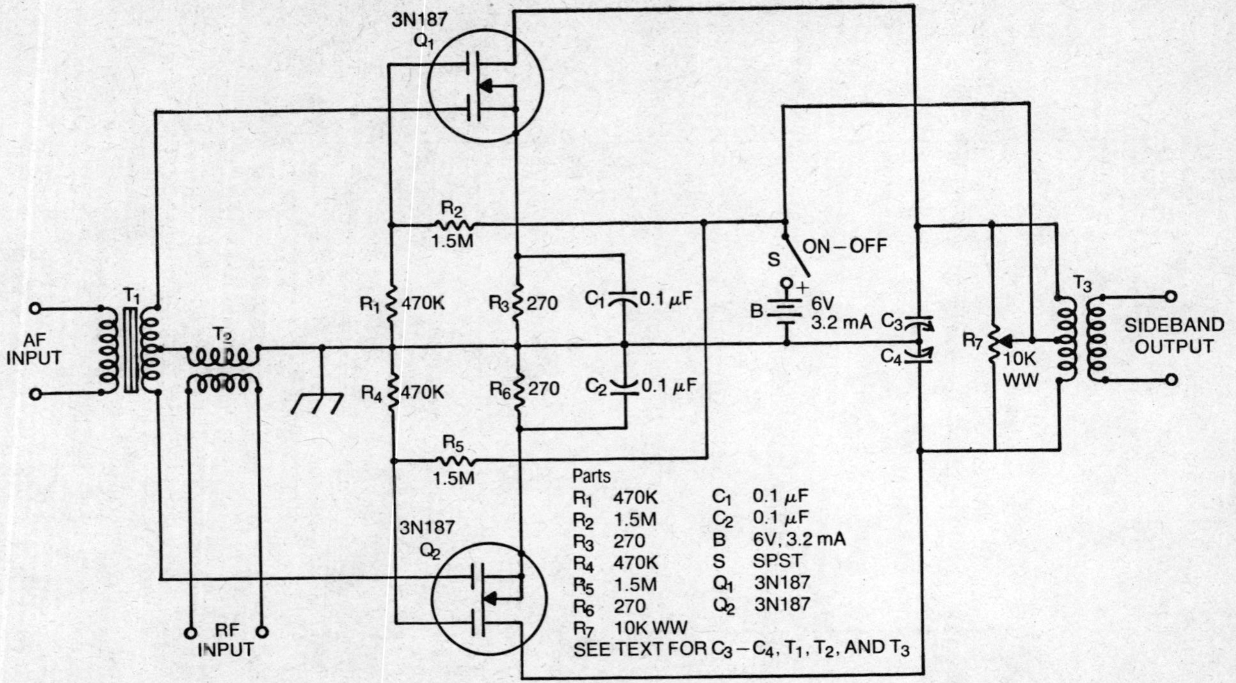

Fig. 7-5. Balanced modulator.

Transformer T_1 must be a high-quality audio component having as accurate a centertap as obtainable; the turns ratio usually is unimportant, but will usually be 1:1 or 2:1. Radio-frequency transformer T_2, like the audio transformer, usually is not critical as to turns ratio. It may be an air-core, ferrite-core, or powdered-iron-core unit. (This unit can be improvised from any IF transformer by disconnecting the internal capacitors, and for high-frequency operation a unity-coupled transformer can be obtained by interwinding two coils of 25 to 50 turns each on a 1 in. diameter form.) Transformer T_3 must be chosen such that its tapped primary resonates with C_3–C_4 at the sideband frequency. Capacitors C_3–C_4 are chosen in the same way.

Some advantage is gained from the use of balanced MOSFETs and matching resistors in the two halves of this circuit. Here, MOSFET Q_1 receives its negative gate-1 bias from the voltage drop resulting from the flow of drain current through source resistor R_3, and its positive gate-2 bias from voltage divider R_1–R_2. Similarly, MOSFET Q_2 receives its negative gate-1 bias from the flow of drain current through source resistor R_6, and its positive gate-2 bias from voltage divider R_4–R_5. With individual MOSFETs some adjustment of resistances R_2 and R_5 may be required for improved balance of the circuit.

Current drain from the 6V supply is approximately 3.2 mA, but this may vary somewhat with individual MOSFETs.

SIMPLE LIGHT-BEAM RECEIVER

Figure 7-6 shows the circuit of a simple receiver for modulated light-beam communications. The pickup device is a high-output, self-generating silicon photocell, *PC* (International S7M-C). The audio output of this cell is presented to gate 1 of the MOSFET through blocking capacitor C_1, which prevents the DC output of the cell from reaching the gate.

The circuit gives excellent results with a high-resistance earpiece. The amplifier is a conventional common-source unit in which gate 1 of the MOSFET receives its negative bias from the voltage drop resulting from the flow of drain current through source resistor R_5, and gate 2 receives its positive bias from voltage divider R_3–R_4. With an individual MOSFET, some increase in maximum obtainable volume may be obtained by experimenting with the value of resistance R_4. Current drain from the 9V supply is approximately 2.6 mA, but this may vary somewhat with individual MOSFETs.

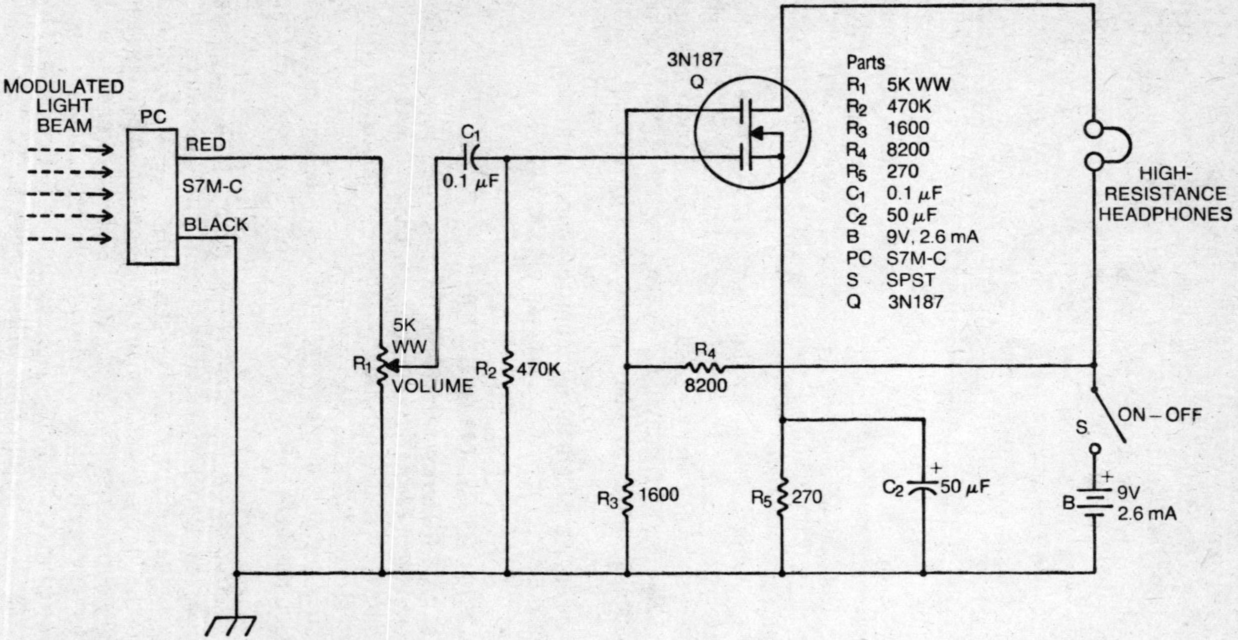

Fig. 7-6 Simple light-beam receiver.

SENSITIVE LIGHT-BEAM RECEIVER

Figure 7-7 shows the circuit of a modulated light-beam receiver having considerably more sensitivity than the simple receiver previously described. It therefore can accommodate a weaker signal than can be handled by the simpler receiver.

In this arrangement, the photocell is followed by a 2-stage MOSFET amplifier having an overall open-circuit voltage gain of approximately 100 when volume control R_7 is set for maximum. The audio output may be applied directly to high-impedance headphones, or the receiver may be followed with an amplifier and speaker. (A suitable external amplifier is described in the section *MOSFET Input for IC Amplifier*, Chapter 2.)

In the receiver circuit, MOSFET Q_1 receives its negative gate-1 bias from the voltage drop resulting from the flow of drain current through source resistor R_5, and its positive gate-2 bias from voltage divider R_3–R_4. Similarly, MOSFET Q_2 receives its negative gate-1 bias from the voltage drop resulting from the flow of drain current through source resistor R_{10}, and its positive gate-2 bias from voltage divider R_8–R_9. With individual MOSFETs, some adjustment of resistances R_4 and R_8 may provide greater sensitivity.

Current drain from the 6V supply is approximately 3.2 mA, but this may vary somewhat with individual MOSFETs.

REGENERATIVE BROADCAST RECEIVER

For the experimenter and hobbyist, regeneration provides a substantial boost in receiver sensitivity without getting into sophisticated circuits. Figure 7-8 shows the circuit of a simple regenerative receiver for the standard broadcast band.

Here, L is a tapped ferrite loopstick antenna rod which allows full coverage of the broadcast band with variable capacitor C_2 (365 pF). The regeneration control is the 25,000Ω rheostat R_4, in the voltage divider for positive gate-2 bias of the MOSFET. Transformer T may be any convenient interstage audio unit having a turns ratio of 1:1, 2:1, or 3:1. The audio output of the receiver is sufficient to drive a high-impedance headset or earpiece connected directly to the output terminals. However, an external amplifier and speaker may be connected to these terminals, if desired (a suitable amplifier is described in the section *MOSFET Input for IC Amplifier*, Chapter 2).

In this circuit, suitable gate-1 negative bias is obtained from the voltage drop resulting from the flow of drain current through source resistor R_3. Current drain from the 9V supply is approximately 2 mA, but this may vary somewhat with

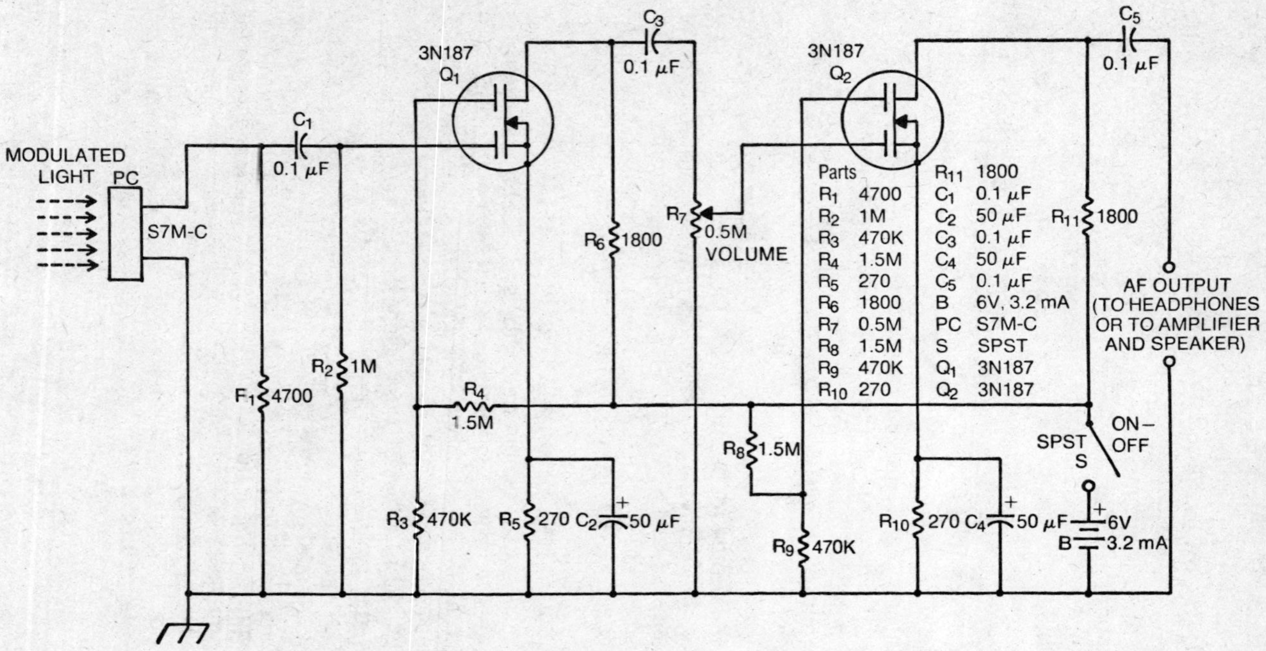

Fig. 7-7. Sensitive light-beam receiver.

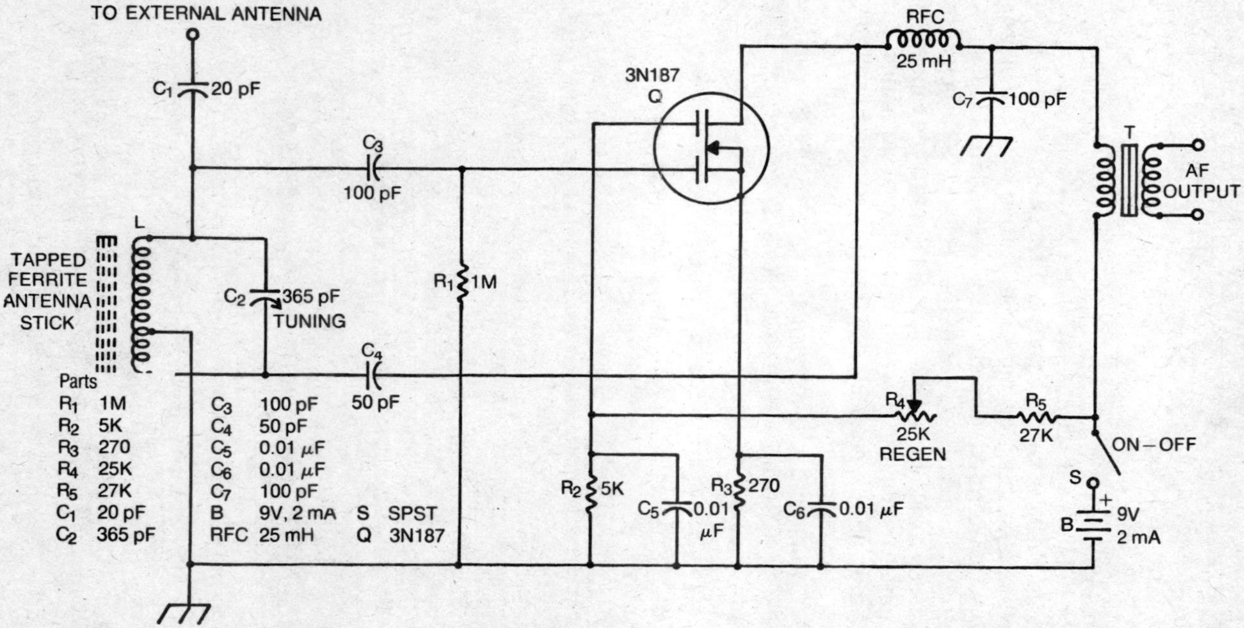

Fig. 7-8. Regenerative broadcast receiver.

individual MOSFETs. The operating constants give the circuit sufficient sensitivity to pick up nearby broadcast stations and high-powered distant stations with only the antenna stick *L*. For more distant stations and weaker ones, an outside antenna and ground must be used.

ALL-WAVE REGENERATIVE RECEIVER

Figure 7-9 shows the circuit of a sensitive regenerative receiver having the general-coverage range of 440 kHz to 30 MHz in five bands. As shown in Table 7-1, five 2-section plug-in coils (L_1–L_2) are used. The circuit is a standard tickler-coil arrangement. The audio output of the receiver is sufficient to drive directly a high-impedance headset or earpiece connected to the output terminals. However, an external amplifier and speaker may also be connected to these terminals (a suitable amplifier is described in the section *MOSFET Input for IC Amplifier*, Chapter 2). Transformer *T* can be any convenient interstage unit having a turns ratio of 1:1, 2:1, or 3:1. The primary of this transformer is RF-bypassed by capacitor C_5.

Table 7-1. Coil-winding data for all-wave regenerative receiver.*

Band	Coil	Winding
Band A 440–1200 kHz	L_1	187 turns No. 32 enameled wire closewound on 1 in. diameter form.
	L_2	45 turns No. 32 enameled wire closewound on same form as L_1. Space 1/16 in. from bottom end of L_1.
Band B 1–3.5 MHz	L_1	65 turns No. 32 enameled wire closewound on 1 in. diameter form.
	L_2	15 turns No. 32 enameled wire closewound on same form as L_1. Space 1/16 in. from bottom end of L_1.
Band C 3.4–9 MHz	L_1	27 turns No. 26 enameled wire closewound on 1 in. diameter form.
	L_2	8 turns No. 26 enameled wire closewound on same form as L_1. Space 1/16 in. from bottom end of L_1.
Band D 8–20 MHz	L_1	10 turns No. 22 enameled wire closewound on 1 in. diameter form.
	L_2	4 turns No. 22 enameled wire closewound on same form as L_1. Space 1/16 in. from bottom end of L_1.
Band E 18–30 MHz	L_1	5½ turns No. 22 enameled wire airwound ½ in. in diameter. Space to winding length of ½ in.
	L_2	4 turns No. 22 enameled wire airwound ½ in. in diameter. Space to winding length of ¼ in. Mount 1/16 in. from bottom end of L_1.

*(Tuning capacitance = 365 pF.)

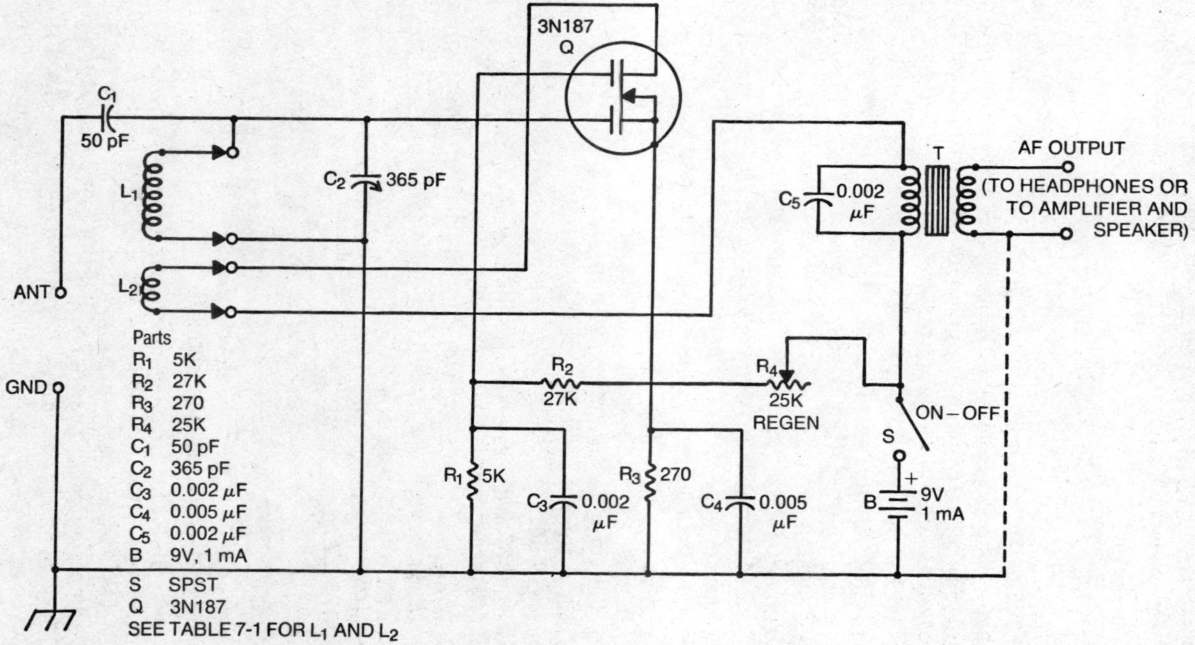

Fig. 7-9. All-wave regenerative receiver.

In this circuit, gate 1 of the MOSFET receives its negative bias from the voltage drop resulting from the flow of drain current through RF-bypassed source resistor R_3. The gate-2 electrode receives its positive bias from voltage divider R_1–R_2–R_4. Current drain from the 9V supply is approximately 1 mA, but this may vary somewhat with individual MOSFETs.

RF MIXER

Figure 7-10 shows the circuit of a simple mixer (converter) stage for radio-frequency applications. Such a stage is readily incorporated into the front end of a superheterodyne receiver or test instrument. Here, the incoming signal (as from RF amplifier stage) is presented to gate 1 of the MOSFET, while the local oscillator signal is applied to gate 2. The coupling capacitances, C_1 and C_2, will be satisfactory in most cases. The IF transformer is selected for the desired intermediate frequency.

In this arrangement, the negative voltage resulting from the flow of drain current through source resistor R_5 is too high for the gate-1 electrode of the MOSFET (the drain current has been selected for the best transconductance, and source resistor R_5 for optimum operation of the MOSFET). To counteract this, a proper positive voltage, developed by voltage divider R_3–R_4 is also applied to gate 1, and the resulting bias has the correct value. Gate 2 of the MOSFET receives its positive bias from voltage divider R_1–R_2. With an individual MOSFET, resistance R_2 in this divider may require some adjustment for maximum conversion gain.

Current drain from the 18V source in the receiver or test instrument is approximately 10 mA, but this may vary somewhat with individual MOSFETs.

TUNING METER

Figure 7-11 shows the circuit of a sensitive tuning meter which may be added to a receiver. The advantage of a MOSFET in this setup is its high input resistance, which results in no discernible loading of the receiver's AGC system.

This circuit is similar to that of an electronic DC voltmeter. The indicating meter is connected in the drain circuit in a 4-arm bridge circuit; the arms of this bridge consist of resistor R_2, the two "halves" of potentiometer R_3, and the internal drain-to-source resistance of the MOSFET. With the receiver detuned from any station, potentiometer R_3 is set to zero the meter. Subsequently, the meter will be deflected upscale

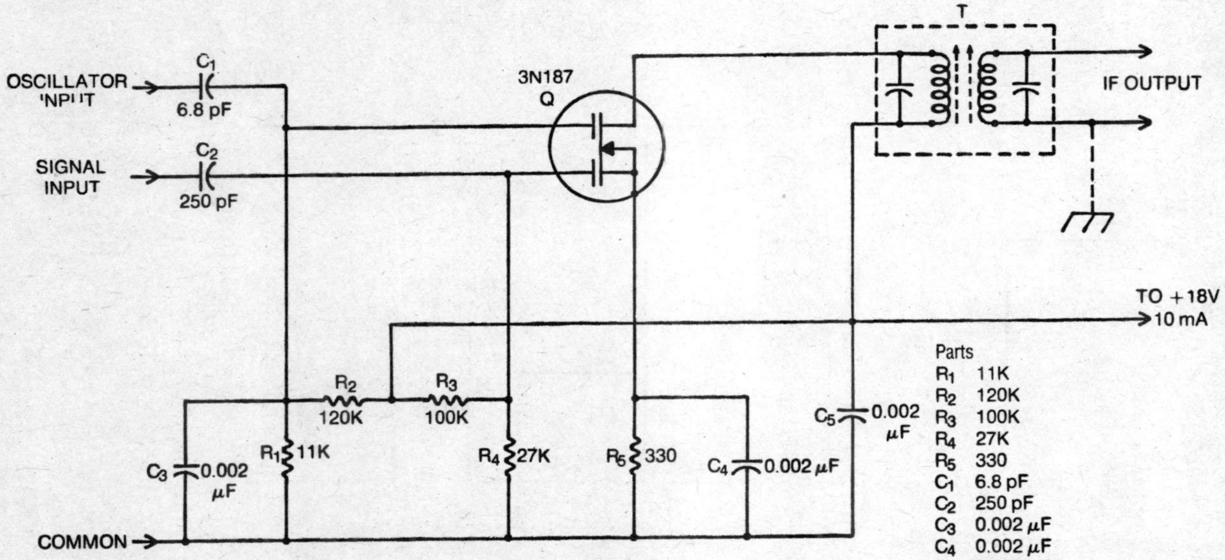

Parts	
R_1	11K
R_2	120K
R_3	100K
R_4	27K
R_5	330
C_1	6.8 pF
C_2	250 pF
C_3	0.002 μF
C_4	0.002 μF
C_5	0.002 μF
Q	3N187

Fig 7-10. Radio-frequency mixer.

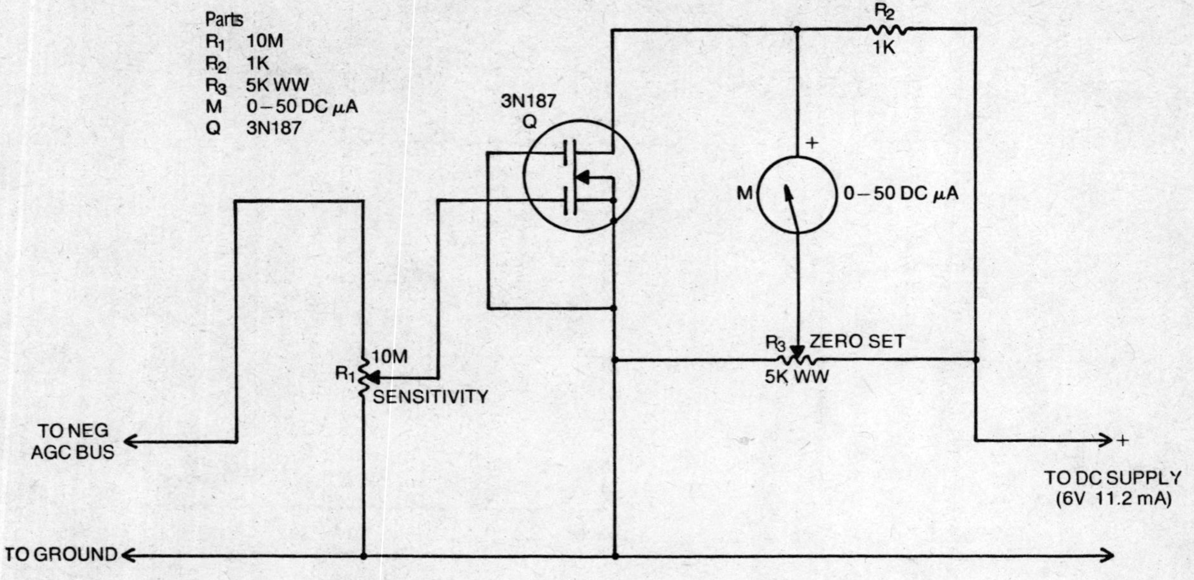

Fig. 7-11. Tuning meter.

whenever a carrier is tuned in, and the deflection will be proportional to the signal strength. In terms of some reference setting of the receiver gain controls, the scale of the meter may be calibrated in microvolts, millivolts, *S*-units, or other indicators of signal strength. For establishing an S-meter scale, make 100 μV equal to S9, 50 μV equal to S8, 25 μV to S7, etc. (Each S-unit should be equal to 6 dB voltage change.)

Current drain from the 6V supply in the receiver is approximately 11.2 mA, but this may vary somewhat with individual MOSFETs.

HETERODYNE ELIMINATOR

The circuit shown in Fig. 7-12 is a highly effective one for eliminating heterodyne whistles in a receiver that has no provision, such as a crystal filter or *Q*-multiplier, for getting rid of this annoyance. This circuit operates in the audio channel of the receiver, which it does not load, owing to the high resistance of R_1. The eliminator may be operated at the AF output of the receiver or patched into the audio channel.

The tuned section of the circuit is a Hall network (C_3–C_4–C_5–R_3–R_4) connected between the input and output stages. The advantage of this network is the tuning it provides with only one potentiometer (R_4). When R_4 is set to the proper point, the network, acting as a bandstop filter, removes the whistle at the corresponding frequency.

The Hall network works best with a low-impedance generator and low-impedance load, and this condition is approximated by driving it with a MOSFET source follower (Q_1) and following it with a bipolar common-emitter stage (Q_2) which exhibits medium-resistance input. A high-impedance headset or earpiece may be connected directly to the output terminals, or an amplifier and speaker may be operated from the output. (A suitable amplifier is described in the section *MOSFET Input for IC Amplifier*, Chapter 2.)

Current drain from the 6V supply is approximately 2.2 mA, but this may vary somewhat with individual MOSFETs and bipolar transistors.

FLEA-POWER "QRP" TRANSMITTER

Figure 7-13 shows the circuit of a low-powered transmitter suitable for clear-channel CW communication and for antenna tuneup operations. It consists of a conventional, keyed crystal oscillator. The DC input power of this transmitter is approximately 60 mW.

In this circuit, the crystal (XTAL) is selected for the desired operating frequency, and the coil set L_1–L_2 is also

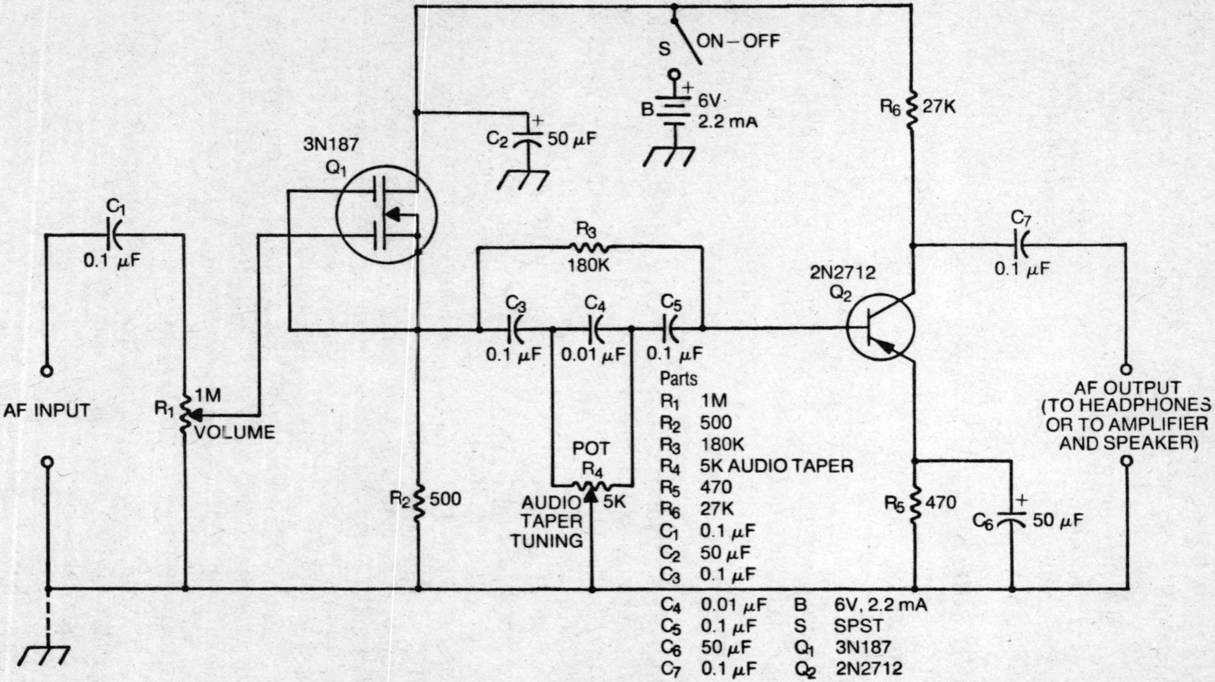

Fig. 7-12. Heterodyne eliminator.

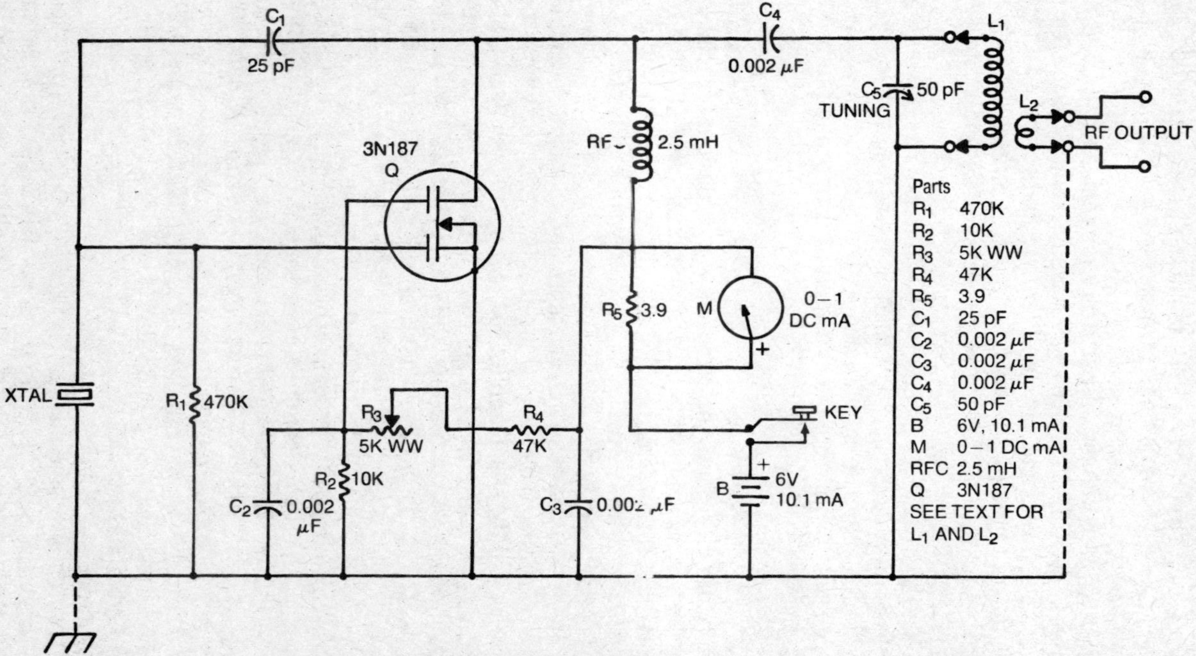

Parts

R_1	470K
R_2	10K
R_3	5K WW
R_4	47K
R_5	3.9
C_1	25 pF
C_2	0.002 μF
C_3	0.002 μF
C_4	0.002 μF
C_5	50 pF
B	6V, 10.1 mA
M	0–1 DC mA
RFC	2.5 mH
Q	3N187

SEE TEXT FOR L_1 AND L_2

Fig. 7-13. Flea-power transmitter

chosen for this frequency. Suitable end-link coil sets for use with the 50 pF tuning capacitor shown are commercially available for the amateur bands. The use of shunt feed in the drain output circuit enables the tank L_1-C_5 to be grounded.

Tuning is simple and straightforward: With the key depressed and no load connected to the output terminals, C_5 is adjusted for minimum dip of drain current as indicated by milliammeter *M*. The load then is connected and adjusted to raise the dipped current to 10 mA. Finally, the tuning is touched up, by readjustment of C_5, for dip. (Some operators, as is well known to the ham reader, prefer to detune slightly to one side of dip.)

The positive bias applied to gate 2 of the MOSFET may be adjusted by means of rheostat R_3. This adjustment permits maximum output to be obtained with an individual MOSFET. The current drain from the 6V supply *B* is approximately 10.1 mA, but this may vary somewhat with individual MOSFETs.

FREQUENCY DOUBLER

Figure 7-14 shows the circuit of a conventional frequency doubler for use in low-powered transmitters and adapted for MOSFET operation. This doubler may be used, for example, in conjunction with the oscillator-type transmitter described in the preceding section.

In this circuit, in the conventional manner of doubler operation, the tank circuit C_4-L_1 is tuned to twice the frequency of the signal applied to the input terminals. Suitable end-link coil sets (L_1-L_2) for use with the 100 pF tuning capacitor shown are commercially available for the amateur bands. The RF output may be either link coupled, as shown by the solid lines, or capacitively coupled by means of C_5 and the dotted lines.

In this circuit, the MOSFET runs at drain-current pinchoff, this condition being obtained by means of proper gate-1 and gate-2 biases. The net negative bias on gate 1 results from the negative voltage drop across source resistor R_5 and the positive output of voltage divider R_1-R_2, while the positive bias on gate 2 results from the output of voltage divider R_3-R_4. Some adjustment of resistances R_2 and R_3 may be required with individual MOSFET. The doubler is fully driven by a 1.5V peak signal at the input terminals.

Tuning is simple and conventional: With switch *S* closed, the driving signal applied to the input terminals, and no load connected to the output terminals, capacitor C_4 is adjusted for minimum dip of drain current as indicated by milliammeter *M*. The load then is connected and adjusted to raise the dipped

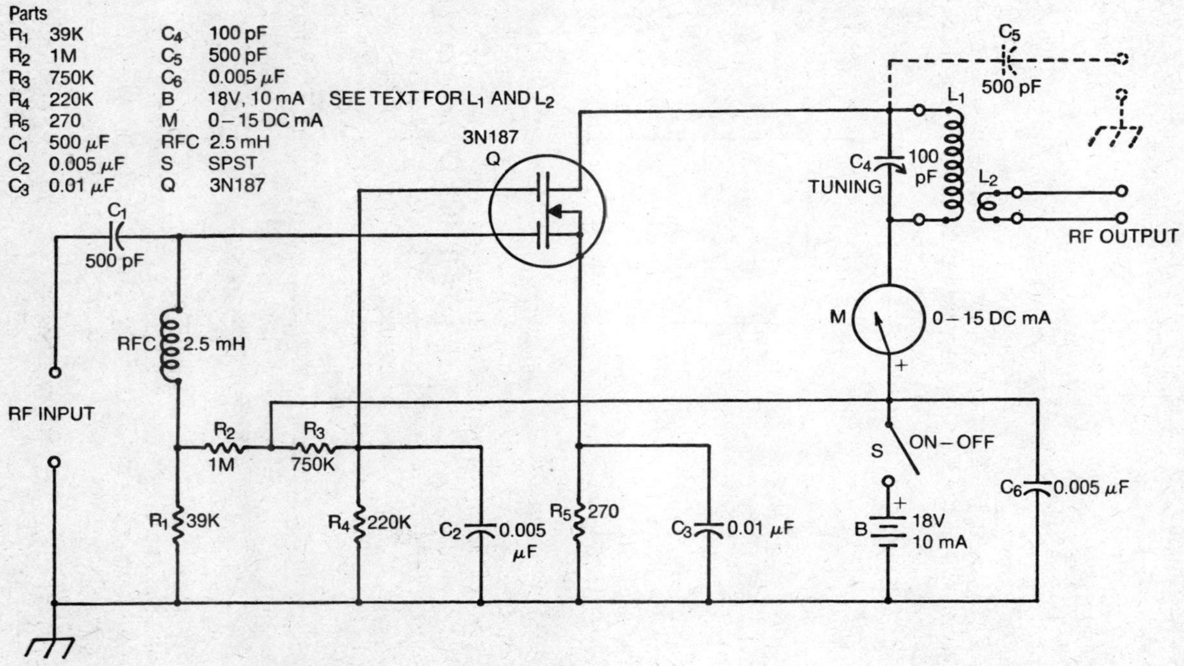

Fig. 7-14. Frequency doubler.

Parts

R_1	1M	C_4	0.01 μF
R_2	470K	C_5	10 μF
R_3	1.5M	C_6	0.005 μF
R_4	270	C_7	0.005 μF
R_5	47K	RFC_1	2.5 mH
C_1	0.01 μF	RFC_2	2.5 mH
C_2	100 pF	S	SPST
C_3	0.002 μF	Q	3N187

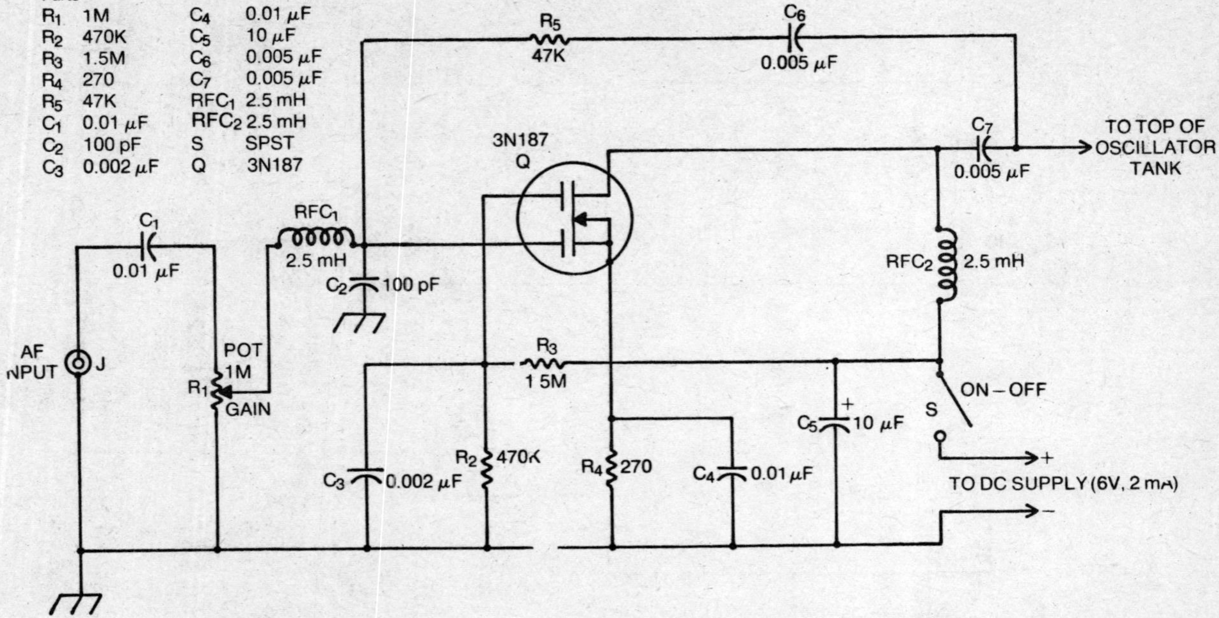

Fig. 7-15 Reactance modulator

current to 10 mA. Finally, the tuning is touched up by readjustment of C_4 for dip. The DC power input of the doubler is approximately 153 mW, but this may vary somewhat with individual MOSFETs.

REACTANCE MODULATOR

Figure 7-15 shows the circuit of a conventional reactance modulator for frequency-modulating a self-excited oscillator. With the drain connected to the tank of the oscillator, this modulator will swing the oscillator frequency proportionate to the amplitude of the audio voltage applied at jack J. Capacitor C_6 and resistor R_5 provide the necessary phase shift to gate 1 of the MOSFET. The entire unit must be mounted close to the oscillator, so that the connection between C_7, C_6, and the oscillator tank may be as short and direct as practicable. This arrangement is suitable not only for FM transmitters, but for sweeping the frequency of a test oscillator.

In this circuit, the negative gate-1 bias is obtained from the voltage drop resulting from the flow of drain current through source resistor R_4, and the positive gate-2 bias from voltage divider R_2–R_3. Current drain from a 6V supply in the transmitter or test instrument is approximately 2 mA, but this may vary with individual MOSFETs.

Index

A

Ac
- null detector 162
- RF relay 83
- voltmeter adapter 157

Active
- high-pass filter 44
- low-pass filter 40, 42
- probe 157

Adapter
- constant-current 170
- output (dual-polarity) 113
- signal tracer 157

AF
- -actuated relay 87
- mixer 31
- oscillator (Colpitts) 120
- oscillator (Franklin) 122
- oscillator (Hartley) 117
- oscillator (phase-shift) 124
- oscillator (sun-powered) 128
- oscillator (tickler feedback) 116
- oscillator (Wein bridge) 126
- oscillator (Villard) 128
- /RF signal tracer 154
- signal tracer adapter 157

AGC amplifier 27

Alarm, carrier-failure 105

All-wave regenerative receiver 179

AM monitor 166

Amplification, gated 35

Amplifier
- bandstop (LC) 52
- current 61
- DC 60
- DC galvanometer 64
- DC voltage 62
- gated-off 38
- gated-on 35
- headphone 56
- high-gain 20
- high-pass (LC-tuned) 44
- IF (double-tuned) 77
- IF (single-tuned) 74
- LC-tuned low-pass 40
- paraphase 33
- peak 46, 49
- peaked response 56
- push-pull RF 71
- RC-coupled 19
- RC-tuned low-pass 42
- regenerative 77
- RF 66
- RF (100 MHz) 71
- RF (transmitter-type) 69
- single-stage 18
- slot (notch) 52, 154
- 2-stage (high-gain) 22
- with AGC 28
- video 80

Amplitude modulation monitor 166

AND gate 105

Audio
- -actuated relay 87
- mixer 31

Audio oscillator
- colpitts 120

Franklin 122

Hartley 117
- phase shift 124
- sun-powered 128
- tickler coil 116
- villard 128
- wien bridge) 126

B

Balanced
- DC voltmeter 147
- modulator 174

Band rejection amplifier 52, 154

Bandstop
- amplifier (LC) 52
- amplifier (RC) 54

Bandwidth, rejection 52

Base connections, MOSFET 15

Bass cut 44

Bistable multivibrator 172

Bridge, Wien 126

Broadcast
- -band coil data 62
- receiver 179

C

Capacitance
- measuring device 151
- relay 98

Carrier-failure alarm 105
Cascade 22
Cathode follower equivalent 25
Center-zero DC voltmeter 149
Channel
- N 11
- P 12

Charge detector 171
Checker, LC (wide range) 157
Code practice monitor 166
Coincidence relay 91
Coil data
- broadcast band 62
- shortwave bands 179

Colpitts
- audio oscillator 120
- oscillator (RF self-excited) 134

Common-source amplifier 18
Constant-current adapter 170
Converter
- impedance (DC) 113
- RF 189
- sine to square 163

Crystal
- -controlled IF oscillator 144
- oscillator (Pierce) 140
- oscillator (standard) 138
- -type frequency standard 152

Current
- amplifier circuit 61
- constant 170

Cutoff, high-frequency 40, 44, 142
Cw monitor 166

D

Dc
- amplifier 60
- galvanometer amplifier 64
- relay (zero current) 81
- signal inverter 110
- source follower 66
- voltage amplifier 62
- voltmeter 145
- voltmeter (balanced) 147
- voltmeter (center zero) 149

Degeneration 22
Delay relay 100
Delayed pull-in relay 100
Depletion mode 12
Demodulator, light beam 177, 179
Detection, null 46
Detector
- charge 171
- null (AC) 162
- proximity 95

Distortion reduction 22
Double-tuned IF amplifier 77
Doubler, frequency 190
Dual
- -gate MOSFET 12
- -input audio mixer 31
- -MOSFET phase inverter 33
- -polarity DC voltmeter 149
- -polarity output adapter 113

E

Earphone amplifier 56
Electroscope 171
Eliminator, heterodyne 187
Emitter follower equivalent 25
Enhancement mode 13

F

Failure, carrier 105
Feedback, negative 22
FET
- junction 11
- vs MOSFET 13

Filter
- high-pass (active) 44
- low-pass (active) 42

Flip-flop 172
Floating gate 17
FM reactance modulator 133
Follower
- source 25
- source (DC) 66

Franklin audio oscillator 122
Frequency
- doubler 190
- standard, self-excited 149

G

Gain, source follower 27
Galvanometer amplifier (DC) 64
Gate
- AND 105
- floating 17
- OR 110
- -protected MOSFET 14

Gated-on amplifier 35
Gated-off amplifier 38
Generator
- audio (tickler-type) 116
- square-wave 163
- tone (Villard) 128
- tone (Wien bridge) 126

H

Hall network 187
Handling care, MOSFET 16
Hartley
- audio oscillator 117
- oscillator 77

Headphone
- amplifier 56
- amplifier (peaked) 56

Heat
- excessive 17
- -operated relay 95

Heterodyne eliminator 187
High-gain amplifier 20
High-pass amplifier 44
Hybrid amp with AGC 27

I

IF amplifier
- double-tuned 77
- input stage 29
- regenerative 77
- single-tuned 74

IF oscillator,
- crystal 144
- self-excited 44

IF/RF signal tracer 155
Impedance converter (DC) 113
Inductance measuring device 162
Input stage for IC amplifier 29
Integrated diodes 14
Interval timer 100
Intrusion relay 98
Inverter
- DC signal 110
- phase 33

J

Junction FET 11
- vs MOSFET 13

L

LC
- checker, wide-range 157
- -tuned AF relay 87
- -tuned amplifier 40
- -tuned high-pass amplifier 44
- -tuned notch amplifier 52
- -tuned peak amplifier 46

Light
- -beam receiver 177
- -operated relay 93
- meter (sensitive) 163

Logic
- AND circuit 105
- OR gate 110

Low
- -pass amplifier 40
- -power transmitter 187

M

Meter
- light 163
- tuning 184

Miniature transmitter 187
Mixer 31
- RF 184

Mode,
- depletion 12
- enhancement 13

Modulation monitor 166
Modulator
- balanced 174
- reactance 193

Monitor
- AM 166
- CW 166

MOSFET
- description 11
- /bipolar amplifier 28
- gate-protected 14
- handling care 16
- input circuit (for IC amp) 30
- mounting 16
- polarities 13
- precautions 16
- structure 12
- symbols 13
- transconductance 13
- vs junction FET 13

Mounting, MOSFET 16
Multiplier, frequency 190
Multivibrator
- bistable 172
- drain-coupled 128

N

N-channel MOSFET 11
Negative feedback 22
Network
- Hall 187
- parallel-T 54

Noise relay 100
Notch
- amplifier (LC) 52
- amplifier (RC) 54

Null
- detection 46
- detector, AC 162
- circuit 51

O

Off gating 38
On gating 35
OR gate 110
Oscillator
- AF (sun-powered) 128
- Colpitts (audio) 120
- crystal (Pierce) 141
- crystal (standard) 138
- /doubler 190
- Franklin (audio) 122
- Hartley 77
- Hartley (audio) 117
- IF (crystal) 144
- IF (self-excited) 144
- multivibrator 128
- RF (Colpitts self-excited) 134
- RF self-excited (Hartley) 134

self-excited RF (tickler) 128
sun-powered (RF) 140
tickler (audio) 116
villard 128
wien bridge 126
Output adapter, dual-polarity 113

P

P-channel MOSFET 12
Parallel-T network 54
Paraphase amplifier 33
Pass
high- 44
low- 40, 42
Peak amplifier 49
Peaked-response amplifier 56
Phase
inverter 33
-shift audio oscillator 124
shifter, stepping 105
shifter, variable 103
Photocell
and relay 93
demodulator 177, 179
Photoelectric relay 93
Photographic light meter 163
Pickup
stray 17
supersonic 174
Pierce crystal oscillator 140
Plate
relay 81
Polarity, MOSFET 13
Positive feedback 77
Probe, active 157
Proximity detector 98
Pulse inverter 110
Push-pull RF amplifier 71

Q

Q-multiplier simulator 187
QRP transmitter 187

R

Reactance modulator 193
Rc
-coupled amplifier circuit 19
-tuned AF oscillator 128
-tuned AF relay 89
-tuned high-pass amplifier 44
-tuned high-pass circuit 46
-tuned low-pass amplifier 42
-tuned notch amplifier 54
-tuned peak amplifier 49
Receiver
al-wave 179
broadcast 179
light beam 177, 179
Reducing distortion 22
Regenerative
IF amplifier 77
receiver (BC) 179
receiver (SW) 179
Rejection bandwidth 52
Relay
AC RF 83
audio-actuated 87
capacitance 98
coincidence 91
heat-operated 95
light-activated 93
noise-operated 100
RF (sensitive) 85
touch-plate 95
zero-current 81
Resistor, source (unbypassed) 22
RF
-actuated relay 83
amplifier, general-purpose 66
amplifier, 100 MHz 71
amplifier, push-pull 71
amplifier (transmitter-type) 69
coils (winding data) 62
colpitts oscillator (self-excited) 134
mixer 184
oscillator, self-excited (tickler) 128
oscillator, sun-powered 140
relay (AC) 83
self-excited oscillator 134
/IF signal tracer 155

S

S-meter 184
Self-excited
colpitts RF oscillator 134
frequency standard 149
Hartley RF oscillator 134
IF oscillator 144
Sensitive
relay 81
RF relay 85
Shielding 17
Shifter
phase 33
phase (variable) 103
Sideband balanced modulator 114
Signal
absorption circuit 49
generator, audio (tickler coil) 116
inverter, DC 110
tracer adapter 157
tracer, AF/RF 154
tracer, IF/RF 155
Silicon wafer 11

Sine-
- to square-wave converter 163
- wave AF oscillator 124

Single-
- ended RF amplifier 70
- gate MOSFET 12
- stage amplifier 18
- tuned IF amplifier 74

Slot
- amplifier (LC) 52
- amplifier (RC) 54

Solar-
- powered audio oscillator 128
- powered RF oscillator 140

Sound switch 100
Source follower 25
- DC 66

Square-wave generator 163
Standard
- crystal oscillator 138
- frequency (self-excited) 140

Static-charge detector 171
Step-type phase shifter 105
Stray pickup 17
Strength meter (RF) 184
Suboscillatory amplifier 77
Sun-powered
- audio oscillator 128
- RF oscillator 140

Superheterodyne converter 184
Supersonic pickup 174
Switch, sound 100

Γ

Tapped-
- capacitance AF oscillator 120
- inductance AF oscillator 117
- capacitance self-excited oscillator 134

Temperature-sensitive relay 95
Thermal relay 45
Tickler-coil oscillator (AF) 116
Timer, interval 100
Time delay relay 100
Tone generator
- colpitts 120
- Hartley 117
- Franklin 122
- tickler-type 116
- villard 128
- wien bridge 126

Touch-plate relay 95
Tracer
- signal (AF/RF) 154
- signal (IF/RF) 155

Transconductance, MOSFET 13
Transmitter
- low-power 187
- RF amplifier 69

Treble cut 42
Tuned RF/IF tracer 155
Tuning meter 184
Two-
- input audio mixer 31
- MOSFET phase inverter 33
- stage amplifier (high-gain) 22

U

Unbypassed source resistor 22

V

Variable phase shifter 103
VHF RF amplifier 71
Video amplifier 50
- circuit 79

Villard audio oscillator 128
Voltage
- amplifier, DC 62
- gain, source follower 27

Voltmeter
- adapter (AC) 157
- booster 50
- DC 145
- DC (center-zero) 149

W

Wafer, silicon 11
Wideband
- amplifier 80
- video amplifier circuit 79

Wien bridge AF oscillator 126
Winding of coils 62

Z

Zeners in FET gates 14
Zero-
- center meter 64